PATTERN DISCOVERY IN
BIOMOLECULAR DATA

PATTERN DISCOVERY IN BIOMOLECULAR DATA

Tools, Techniques, and Applications

Edited by
JASON T. L. WANG
BRUCE A. SHAPIRO
DENNIS SHASHA

CABRINI COLLEGE LIBRARY
610 King of Prussia Road
Radnor, PA 19087

New York   Oxford
Oxford University Press
1999

3905/708

Oxford University Press

Oxford   New York
Athens   Auckland   Bangkok   Bogotá   Buenos Aires   Calcutta
Cape Town   Chennai   Dar es Salaam   Delhi   Florence   Hong Kong   Istanbul
Karachi   Kuala Lumpur   Madrid   Melbourne   Mexico City   Mumbai
Nairobi   Paris   São Paulo   Singapore   Taipei   Tokyo   Toronto   Warsaw

and associated companies in
Berlin   Ibadan

Copyright © 1999 by Oxford University Press, Inc.

Published by Oxford University Press, Inc.
198 Madison Avenue, New York, New York 10016

Oxford is a registered trademark of Oxford University Press.

All rights reserved. No part of this publication may be reproduced,
stored in a retrieval system, or transmitted, in any form or by any means,
electronic, mechanical, photocopying, recording, or otherwise,
without the prior permission of Oxford University Press.

Library of Congress Cataloging-in-Publication Data
Pattern discovery in biomolecular data : tools, techniques, and applications /
    edited by Jason T. L. Wang, Bruce A. Shapiro, and Dennis Shasha.
        p.   cm.
    Includes bibliographical references and index.
    ISBN 0-19-511940-1
    1. Nucleotide sequence—Data processing.   2. Aminoacid sequence—
Data processing.   3. Pattern recognition systems.   4. Information storage
and retrieval systems—Nucleotide sequence.   I. Wang, Jason T. L.
II. Shapiro, Bruce A.   III. Shasha, Dennis Elliott.
QP620.P38   1999
572.8'5'0285—dc21   98-18934

9 8 7 6 5 4 3 2

Printed in the United States of America
on acid-free paper

To our families

Jason: Bao-Jei, Tzu-Chen Yen,
        Lynn, and Tiffany

Bruce: Judy,
        Stacy, and Rachell

Dennis: Alfred, Hanina,
        Carol, Joe, Jeff, Ariana, Nick,
        Robert, Ellen, Jordan, David,
        Karen, Cloe, and Tyler

Contents

Part II. Finding Patterns in 3D Structures

Part III. System Components for Discovery

Contributors

Timothy L. Bailey
San Diego Supercomputer Center
P.O. Box 85608
San Diego, CA 92186

Michael E. Baker
Department of Medicine
University of California
San Diego, CA 92093

Andrea Califano
IBM T.J. Watson Research Center
P.O. Box 704
Yorktown Heights, NY 10598

Gung-Wei Chirn
Bioinformatics Pre-Clinical R&D
Hoffmann-La Roche Inc.
Nutley, NJ 07110

Diane J. Cook
Department of Computer Science
   and Engineering
University of Texas
Arlington, TX 76019

Kathleen Currey
National Cancer Institute
Frederick, MD 21702
   and
Medical School
University of Maryland
Baltimore, MD 21201

Judith Bayard Cushing
The Evergreen State College
Olympia, WA 98505

Charles P. Elkan
Department of Computer Science
   and Engineering
University of California
San Diego, CA 92093

Suzanne Fortier
Departments of Chemistry and
   Computing Science
Queen's University
Kingston, Canada, K7L 3N6

Gehad Galal
Department of Computer Science
   and Engineering
University of Texas
Arlington, TX 76019

Janice Glasgow
Department of Computing and
   Information Science
Queen's University
Kingston, Canada, K7L 3N6

William N. Grundy
Department of Computer Science
   and Engineering
University of California
San Diego, CA 92093

Jorja G. Henikoff
Fred Hutchinson Cancer
    Research Center
P.O. Box 19024
Seattle, WA 98109

Lawrence B. Holder
Department of Computer Science
    and Engineering
University of Texas
Arlington, TX 76019

Tim Hunkapiller
Department of Molecular
    Biotechnology
University of Washington
Seattle, WA 98195

Minoru Kanehisa
Institute for Chemical Research
Kyoto University
Uji, Kyoto 611, Japan

Wojciech Kasprzak
Intramural Research Support
SAIC Frederick
National Cancer Institute
Frederick, MD 21702

Elizabeth Kutter
The Evergreen State College
Olympia, WA 98505

Justin Laird
The Evergreen State College
Olympia, WA 98505

Bin Li
Department of
    Computer Science
Courant Institute of
    Mathematical Sciences
New York University
New York, NY 10012

Thomas G. Marr
Department of Molecular, Cellular
    and Developmental Biology
University of Colorado
    and
Genomica Corporation
4001 Discovery Drive
Boulder, CO 80303

Aleksandar Milosavljević
Genometrix Inc.
3608 Research Forest Drive
The Woodlands, TX 77381

Daniel Platt
IBM T.J. Watson Research Center
P.O. Box 704
Yorktown Heights, NY 10598

Isidore Rigoutsos
IBM T.J. Watson Research Center
P.O. Box 704
Yorktown Heights, NY 10598

Steve Rozen
Center for Genome Research
Whitehead Institute for
    Biomedical Research, MIT
Cambridge, MA 02139

Bruce A. Shapiro
Laboratory of Experimental and
    Computational Biology
National Cancer Institute
Frederick, MD 21702

Dennis Shasha
Courant Institute of
    Mathematical Sciences
New York University
New York, NY 10012

David Silverman
IBM T.J. Watson Research Center
P.O. Box 704
Yorktown Heights, NY 10598

Evan Steeg
Department of Computing and
    Information Science
Queen's University
Kingston, Canada, K7L 3N6

Kentaro Tomii
Institute for Chemical Research
Kyoto University
Uji, Kyoto 611, Japan

Jason T. L. Wang
Department of Computer and
    Information Science
New Jersey Institute of
    Technology
Newark, NJ 07102

Zhiyuan Wang
Department of Computer and
    Information Science
New Jersey Institute of
    Technology
Newark, NJ 07102

Jin Chu Wu
Intramural Research Support
SAIC Frederick
National Cancer Institute
Frederick, MD 21702

David P. Yee
Biomolecular Informatics
Zymogenetics Inc.
Seattle, WA 98102

Kaizhong Zhang
Department of Computer Science
University of Western Ontario
London, Canada, N6A 5B7

Frank Zucker
Department of Molecular
    Biotechnology
University of Washington
Seattle, WA 98195

Introduction

Background and Goal

Biomolecular data include DNA (deoxyribonucleic acid), RNA (ribonucleic acid), protein sequences, and their two-dimensional (2D) and three-dimensional (3D) structures. DNA acts like a biological computer program that spells out the instructions for making proteins. It has a twisted double helical structure. Each strand of the DNA double helix is a polymer built from four components, called *nucleotides*: A, T, C, and G (the abbreviations for adenine, thymine, cytosine, and guanine). The two strands of DNA are complementary: whenever there is a T on one strand, there is an A in the corresponding position on the other strand; whenever there is a G on one strand, there is a C in the corresponding position on the other. DNA can be represented by a sequence of these four letters, or *bases*.

Like DNA, RNA is a long molecule but is usually single stranded, except when it folds back on itself. It differs chemically from DNA by containing ribose sugar instead of deoxyribose and containing the base uracil (U) instead of thymine. Thus, the four bases in RNA are A, C, G, and U.

A protein is also a polymer, constructed by hundreds or thousands of amino acids. The most popular representation model for biologists to describe a protein is to use the sequence. A protein sequence is made up of 20 amino acids, each represented by a letter: alanine (A), cysteine (C), aspartic acid (D), glutamic acid (E), phenylalanine (F), glycine (G), histidine (H), isoleucine (I), lysine (K), leucine (L), methionine (M), asparagine (N), proline (P), glutamine (Q), arginine (R), serine (S), threonine (T), valine (V), tryptophan (W), and tyrosine (Y).

With the significant growth of the amount of biomolecular data, it becomes increasingly important to develop new techniques for extracting "knowledge" from the data. *Pattern discovery* is a fundamental operation in such a domain. A pattern here has a broad sense, which may refer to repeatedly occurring "words," or substrings, in a genomic sequence; blocks of conserved segments in a group of functionally related protein sequences; common motifs in RNA and protein secondary structures; or recurrent 3D structural motifs in the polymers. Such patterns have many applications in, for example, the detection of genetic diseases, the classification of DNA sequences, the prediction of RNA and protein secondary and tertiary structures, the investigation of structure-function relationships, the understanding of protein evolution, and rational drug design.

The aim of this book is to introduce you to some of the best techniques of pattern discovery in molecular biology in the hope that you will build on them to make new discoveries on your own. The techniques draw from many fields of mathematical science ranging from graph theory to informa-

Table I.1   Data types addressed in each chapter.

Chapter	DNA/RNA sequence	Protein sequence	2D graph	RNA structure	3D molecule
1	×				
2		×			
3		×			
4	×	×			
5		×			×
6		×	.		×
7					×
8	×	×		×	
9	×		×		
10	×	×			
11	×			×	

tion theory to statistics to computer vision. We hope you find the book as fascinating to read as we have found it to write and edit.

Organization of the Book

A wonderful and at the same time frustrating feature of biological pattern discovery is that there is no best method for finding patterns. The fundamental reason is ontological: many representations of the data are possible (e.g., sequence, tree, graph, energy map), and many forms of patterns are possible within each representation (e.g., alignment, regular expression, probability matrix). For this reason, complementary discovery methods often outperform single methods. Our approach in this book is to present methods in their purest form. You can then choose the method or combination of methods that best fits your applications. Below we describe in more detail the approaches in each chapter that follows. Table I.1 lists the data types addressed in each chapter, and table I.2 summarizes the techniques used in each chapter.

Finding Patterns in Sequences

Part I addresses the basics of finding sequence patterns. Starting from the simple idea that biological sequences are messages to be decoded, Aleksandar Milosavljević explains in chapter 1 how to discover inherent similarities in sequences and to correct for probe biases. The author goes on to explore an information theoretic criterion for life itself and applies it to Alu evolution. Other applications will suggest themselves to the reader. For example, the author addresses a problem of finding "simple" DNA regions.

Table I.2 Techniques used in each chapter.

Technique	Chapter										
	1	2	3	4	5	6	7	8	9	10	11
Clustering					×	×	×			×	×
Hashing			×				×				
Visualization											×
Genetic algorithm											×
Graph matching					×		×		×		
Graph traversal		×						×			
Knowledge base					×				×		
Machine learning			×		×						
MDL principle	×								×		
Sequence alignment						×					×
Statistics/probability	×		×	×							

A region is simple if it contains exactly repeated occurrences of a few words. Such regions are informative for genetic mapping and diseases. The author employs the minimum description (or encoding) length principle, Kolmogorov complexity, and data compression techniques to approach the problem. He also discusses applications of the methods to the problems of comparing DNA sequences of arbitrary complexity and DNA sequence recognition by hybridization.

In chapter 2, Jorja Henikoff describes a system, called Protomat Block Maker and Searcher, for discovering "blocks" from a group of functionally related proteins. Here, the blocks consist of multiply aligned nongapped sequence segments that correspond to the most highly conserved regions of the given group of sequences. The system works by enumerating candidate sets of blocks using a depth first search algorithm. It then scores the blocks and selects the highest scoring set. The resulting blocks can be used to classify sequences of unknown function into known protein families. The Protomat system is public domain software accessible via e-mail as well as World Wide Web (WWW) browsers.

In chapter 3, Timothy Bailey, Michael Baker, Charles Elkan, and William Grundy present a learning tool using artificial intelligence and expectation maximization techniques, called MEME, for finding patterns or motifs shared by a set of proteins. The patterns are used in conjunction with two other software tools, MAST and Meta-MEME, a hidden Markov-based method, to search protein databases for distantly related homologs (proteins with common ancestors). A parallel version of MEME is accessible on the Web, maintained at the San Diego Supercomputing Center.

Jason Wang and coauthors propose in chapter 4 a sampling technique to discover patterns consisting of simple regular expressions, that is, patterns composed of segments, separated by arbitrarily long gaps, that occur frequently in the given set of sequences. The authors apply the patterns, together with a hash-based fingerprint method, to conducting DNA and protein sequence classification. Like Protomat and MEME, this software is accessible on the Web.

Finding Patterns in 3D Structures

Janice Glasgow, Evan Steeg, and Suzanne Fortier begin part II by describing in chapter 5 four types of patterns occurring in protein sequences as well as secondary and tertiary structures. The authors review various approaches, based on clustering and graph matching, for discovering the motifs. This survey shows how the different techniques may be complementary.

In chapter 6, Kentaro Tomii and Minoru Kanehisa propose an approach to the 3D motif discovery problem based on converting 3D protein coordinate data into symbol strings. The authors then detect structural motifs by using a sequence alignment technique to compare the symbol strings. These efforts contribute to the investigation of the relationship between protein tertiary structure and the corresponding amino acid sequence.

In chapter 7, Isidore Rigoutsos, Daniel Platt, Andrea Califano, and David Silverman apply a technique from computer vision, known as geometric hashing, to discover patterns in 3D structures. The basic idea of geometric hashing is to cover a large object by small, possibly overlapping pieces. To match two large objects, one looks for matches among the small pieces and then tries to find collections of matches that are mutually consistent. The authors combine the technique with a clustering method to find molecules in a database that exhibit, in the presence of rotatable bonds, the maximum structural overlap with a given query molecule.

System Components for Discovery

Since most pattern discovery techniques are computationally expensive, improving their run-time performance becomes important. Part III begins with chapter 8, where Bin Li, Dennis Shasha, and Jason Wang present a general framework for pattern discovery in a parallel environment consisting of occasionally idle workstations. The authors model the discovery task as traversing a directed acyclic graph and invent intelligent heuristics to prune unnecessary traversal.

In chapter 9, Diane Cook, Lawrence Holder, and Gehad Galal describe a system, called Subdue, for finding repetitive substructures in 2D graphs. Like Milosavljević's method, Cook et al.'s system is based on the minimum description length (MDL) principle, but it employs an inexact graph matching algorithm and background knowledge during the discovery process. The authors propose an algorithm to partition a graph for multiple processors. Their experimental results indicate excellent performance when running the algorithm in both parallel and distributed environments.

In chapter 10, David Yee and coauthors present a system, called Overview, that tracks and manages inputs and results from sequence comparison programs. The system provides the capability of organizing these results into "clusters" that can be marked, named, annotated, and manipulated. Objects in clusters can themselves be sorted, filtered, and annotated. A prototype of the system has been implemented in Smalltalk.

In chapter 11, Bruce Shapiro, Wojciech Kasprzak, Jin Chu Wu, and Kathleen Currey describe a family of software and algorithms to support pattern matching and pattern discovery efforts at the National Cancer Institute. In particular, the authors concentrate on the discussion of a parallel genetic algorithm for RNA folding, and the development of an RNA structure analysis workbench, Structurelab, that is used to analyze and integrate RNA structural data from many sources. Graphics utilities are employed to help visualize RNA structures.

Support on the World Wide Web

This book's homepage is

`http://www.cis.njit.edu/~jason/publications/biopat.html`

This page provides the up-to-date information and errors found in the book. It also provides links to biological pattern discovery tools and some major biological database centers around the world.

Acknowledgments

The book is the result of a two-year effort. We thank the contributing authors for meeting the stringent deadlines and for helping to compile and define the terms in the glossary. We also thank many of our current and former students as well as colleagues for useful discussions while preparing this book, especially Nikolaos Bourbakis, Sitaram Dikshitulu, James Gattiker, Wen-Lian Hsu, Qicheng Ma, James McHugh, Takenao Ohkawa, Song Peng, Gautam Singh, Abdullah Tansel, Carol Venanzi, Paul Wang, Xiong Wang, Cathy Wu, and Maisheng Yin.

The U.S. National Science Foundation and U.S. National Institutes of Health have generously supported this interdisciplinary field in general and much of the work presented here in particular.

Kirk Jensen at Oxford University Press was a wonderfully supportive executive editor, obtaining helpful reviews, and giving advice on presentation and approach. His assistants, Gabriel Alkon and Jael Wagener, were there whenever problems cropped up. Finally, a special thanks goes to our production editor Lisa Stallings for her thoughtful comments on drafts of the book that improved its format and content. We are to blame for any remaining problems.

Part I. Finding Patterns in Sequences

Chapter 1

Discovering Patterns in DNA Sequences by the Algorithmic Significance Method

Aleksandar Milosavljević

The parsimony method for reconstruction of evolutionary trees (Sober, 1988) and the minimal edit distance method for DNA sequence alignments (e.g., Waterman, 1984) are both based on the principle of Occam's Razor (e.g., Losee, 1980; also known as the Parsimony principle). This principle favors the most concise theories among the multitudes that can possibly explain observed data. The conciseness may be measured by the number of postulated mutations within an evolutionary tree, by the number of edit operations that transform one DNA sequence into the other, or by another implicit or explicit criterion. A very general mathematical formulation of Occam's Razor has been proposed via minimal length encoding by computer programs (for recent reviews, see Cover and Thomas, 1991; Li and Vitányi, 1993).

Algorithmic significance is a general method for pattern discovery based on Occam's Razor. The method measures parsimony in terms of encoding length, in bits, of the observed data. Patterns are defined as datasets that can be concisely encoded. The method is not limited to any particular class of patterns; the class of patterns is determined by specifying an encoding scheme.

To illustrate the method, consider the following unusual discovery experiment:

1. Pick a simple pseudorandom generator for digits from the set {0, 1, 2, 3}.
2. Pick a seed value for the generator and run it to obtain a sequence of 1000 digits; convert the digits to a DNA sequence by replacing all occurrences of digit 0 by letter A, 1 by G, 2 by C, and 3 by T.

3. Submit the sequence to a similarity search against a database containing a completely sequenced genome of a particular organism.

Assume that after an unspecified number of iterations of the three steps, with each iteration involving a different random generator or seed value or both, the search in the third step finally results in a genomic sequence highly similar to the query sequence.

Does the genomic sequence contain a pattern? To argue for the presence of a pattern, one may directly apply the algorithmic significance method. First, a typical simple pseudorandom generator is a short program, say x bits long, including the encoding of the seed value. To specify the genomic sequence, one would need to specify the differences (the term "mutation" does not apply here!) from the query sequence using an additional y bits and the position of the similarity region within the genome using an additional z bits. Thus, the genomic sequence can be *algorithmically* encoded in $x + y + z$ bits. On the other hand, the sequence can also be encoded letter by letter, using two bits per letter, say, in a total of 2000 bits. Assuming that the sum $x + y + z$ is d bits less than the 2000 bits required to specify the sequence letter by letter, theorem 1.1 (see section 1.3) states that the genomic sequence contains a pattern at the significance level 2^{-d}.

It should be emphasized that in the above example no assumptions were made about the random generator in step 1, except that it is encoded by a short program. Also, no assumptions were made about the number of sequences that can possibly be generated by the pseudorandom generator, except that the particular seed value can be encoded in few bits. The sole basis for assigning the significance value is the existence of a concise algorithmic description of the genomic sequence.

Thus, paradoxically, any concise pseudorandom generator generates nonrandom sequences. In fact, the generator itself defines a class of patterns. The above example is unusual in that the class of patterns defined by random generators is not expected to be observed in nature.

In this chapter I review four applications of the algorithmic significance method. All of the applications have been previously reported in separate articles; here I provide an updated and unified review. The first three applications are practical, while the fourth is only conceptual.

1. *Application 1: discovering simple patterns.* The basic version of the method is applied to discover simple repetitive patterns in DNA sequences (Milosavljević and Jurka, 1993a). Patterns are defined as regions that can be encoded concisely.

2. *Application 2: discovering similarity.* An extended version of the method allows comparison of DNA sequences of arbitrary complexity (Milosavljević, 1995a). The essence of the extended method is to measure similarity by computing the difference between the sum of individual encoding lengths of two compared sequences on one side and their joint encoding length on the other side.

3. *Application 3: sequence recognition by hybridization.* DNA sequences that are only partially determined by hybridization experiments are compared against known full sequences to recognize similarity (Milosavljević, 1995c; Milosavljević et al., 1996b).

4. *Application 4: discovering global patterns of life.* The extended method is applied as a test for detecting life, according to Chaitin's general mathematical definition (Milosavljević, 1995a).

In the following sections I first present the necessary background: the encoding schemes that are used in the applications, an algorithm for computing minimal encoding lengths, and the algorithmic significance theorems. After presenting the background, I review the four applications.

1.1 Encoding Schemes

There are a large variety of encoding schemes, well beyond the usual edit operations (insertions, deletions, and point mutations), that can be used to encode a target sequence t relative to a source sequence s. Let A be a particular encoding scheme and let $I_A(t|s)$ be the number of bits needed to encode target t relative to source s by using the encoding scheme A. A short encoding length indicates similarity of source and target according to scheme A.

Alternatively, the target sequence t may be encoded by itself (without reference to source s) using an appropriate encoding scheme A. $I_A(t)$ denotes the encoding length. Occurrences of patterns are defined as regions that are compressible according to A. An encoding scheme A may be designed for detecting repetitive or other kinds of patterns.

In this section I survey three encoding schemes used in the four applications. All three schemes are variations of the Original Pointer Textual Substitution data compression scheme (Storer, 1988).

1.1.1 Scheme 1: Simple DNA Sequences

Sequence t is defined to be *simple* if it can be encoded by using a small number of bits. To be more precise, let $I_A(t)$ denote the encoding length of t, in bits. Assume the encoding scheme A where occurrences of some words in t are replaced by pointers to the previous occurrences of the same words in t. Assume that a pointer consists of two positive integers: the first integer indicates the beginning position of a previous occurrence of the word within t, while the second integer indicates the length of that word. For example, sequence

AGTCAGTTTT

may be encoded as

AGTC(1,3)(7,3).

The decoding algorithm consists of the following two steps:

1. Replace each pointer by a sequence of pointers to individual letters, and
2. Replace the new pointers by their targets in the left-to-right order.

Continuing the example, the first step would yield

AGTC(1,1)(2,1)(3,1)(7,1)(8,1)(9,1),

and the second step would yield the original sequence. From this decoding algorithm it should be obvious that the original sequence can be obtained despite overlaps of pointers and their targets, as is the case with pointer $(7, 3)$ in this example.

Since the goal is to encode concisely, it pays off to replace a word by a pointer only if the letter-by-letter encoding of the same word is longer. To measure the encoding lengths precisely, we now assume the following encoding scheme for letters and pointers: each element (either a letter or a pointer) is encoded by a log 5-bit (where log denotes logarithm base 2) preamble that either specifies a letter or announces a pointer. In addition, a pointer also contains the $2 \log n$-bit encoding of the two positive integers that do not exceed the length n of the sequence. Thus, the total encoding length of a pointer is $\log 5 + 2 \log n$. Based on this encoding scheme, it pays off to replace an occurrence of a word of length k by a pointer to its previous occurrence if $k * \log 5 > \log 5 + 2 * \log n$. In practical applications, the pointer size can be smaller due to encoding of relative positions and lengths and to other improvements in the encoding scheme.

An encoding of a sequence may be conveniently represented by inserting dashes between letters and pointers and by replacing pointers by their targets. The encoding of our example would then be represented by

A-G-T-C-AGT-TTT.

1.1.2   Scheme 2: Encoding a Target DNA Sequence Relative to a Source Sequence

We now consider the case where one sequence is encoded relative to the other using long words that the two sequences have in common. More precisely, a target sequence t is encoded in $I(t|s)$ bits by replacing some of its words by pointers to the occurrences of the same words in the source sequence s.

Consider an example where the target sequence is

GATTACCGATGAGCTAAT

and the source sequence is

ATTACATGAGCATAAT.

The occurrences of some words in the target may be replaced by pointers indicating the beginning and the length of the occurrences of the same words in the source, for example,

G(1,4)CCG(6,6)(13,4).

As in scheme 1, one may then represent the encoding of a sequence by inserting dashes to indicate the parsing, for example,

G-ATTA-C-C-G-ATGAGC-TAAT.

The exact number of bits needed to encode letters and pointers is calculated similarly as in scheme 1. The only significant difference is that the pointers now point to the occurrences of words in the source sequence s.

1.1.3 Scheme 3: Encoding a Target Sequence Relative to a List of Words

So far we have assumed that the source sequence s is known completely. Now consider the case where the source s is represented by a possibly incomplete list of words. Such lists are typically obtained by hybridization experiments where a DNA fragment is queried by short oligomer probes (Drmanac and Drmanac, 1994); a positive hybridization with a particular oligomer probe indicates presence of a complementary oligomer, that is, a word, in the DNA fragment.

Hybridization experiments typically detect words with a certain degree of error. As an example, consider the following four words that are detected with two errors (indicated by carets):

```
1          GAAGTTGC
2              TTGCGCAT
3                  GTATGCAC
4                ^    CCACAAGT
                 ^
```

GAAGTTGCGCATGCACAAGT

Note that if the words cover the sequence completely and without error and if the overlaps between the words are long enough, then the sequence can be completely reconstructed by a simple algorithm—this is the basic idea of sequencing by hybridization (Drmanac et al., 1989, 1993).

An encoding scheme that takes advantage of possible word overlaps can be designed. A pointer now consists of three numbers: an index of a word, the beginning position within the word, and length. The most important point is that the length can exceed the length of an individual word, provided a unique continuation can be found based on the word suffix (this is how word overlaps may reduce the number of pointers). The sequence from the above example can be encoded relative to the four words above by a single pointer:

$$(1,1,20)$$

The crucial feature of the new decoding algorithm is that it recognizes unique continuations of a word by finding its longest suffix that has a unique additional occurrence. More precisely, if the length recorded in a pointer exceeds the word length, the decoding algorithm breaks down the pointer into two new pointers. The first new pointer is the same as the original pointer with the length decreased so as not to exceed the length of the word. The second new pointer points to the word that contains an occurrence of the longest suffix of the original word; it is required that the occurrence be unique and not itself be a suffix. This procedure is repeated as long as there are pointers that point beyond a word. In our example, the suffix TTGC of word 1 points to word 2, resulting in the pointer being broken down into the following two:

$$(1,1,8)(2,5,12)$$

After repeating the procedure two more times using the suffixes AT and CAC, we get the following list of pointers that point within individual words:

$$(1,1,8)(2,5,4)(3,5,4)(4,5,4)$$

The next step in the decoding algorithm is a simple replacement of pointers by targets, as described above, yielding the correct target sequence.

It is important to note that this coding scheme implicitly takes advantage of overlaps that are not necessarily exact, for example, overlaps between words 2 and 3 (mismatch in the second position of word 3) and between 3 and 4 (mismatch in the first position of word 4). Hybridization errors do not hinder encoding as long as the subword structure provides enough information about continuations. Indeed, the correct target sequence would be encoded by the same single pointer even in the absence of hybridization errors.

To factor out the effect of possible repetitive patterns, the scheme described in this subsection is in practice combined with scheme 1: a pointer can point either to the word list s or to the target t, depending on which results in shorter encoding.

1.2 Algorithms for Minimal Length Encoding

We here present the algorithms for computing minimal encoding lengths for schemes 1, 2, and 3. The algorithms are similar, reflecting the similarity of the encoding schemes. I first describe the algorithm that computes $I_A(t|s)$ for scheme 2. The input to the algorithm is target sequence t, source sequence s, and the encoding length $p \geq 1$ of a pointer. Since it is only the ratio between the pointer length and the encoding length of a letter that matters, the two values are linearly scaled so that the encoding length of a letter becomes 1.

Let n be the length of sequence t and let t_k denote the $(n-k+1)$-letter suffix of t that starts in the kth position. Using suffix notation, we write t_1 instead of t. $I(t_k|s)$ denotes the minimal encoding length of the suffix t_k relative to source s. Finally, let $l(i)$, where $1 \leq i \leq n$, denote the length of the longest word that starts at the ith position in target t and that also occurs in the source s. If the letter at position i does not occur in the source, then $l(i) = 0$. Using this notation, we may now state the main recurrence:

$$I(t_i|s) = \min(1 + I(t_{i+1}|s), p + I(t_{i+l(i)}|s))$$

Proof of this recurrence can be found in Storer (1988).

Based on this recurrence, the minimal encoding length can now be computed in linear time by the following two-step algorithm:

1. Compute values $l(i)$, $1 \leq i \leq n$, in linear time by using a directed acyclic word graph data structure that contains the source s (Blumer et al., 1987).
2. Compute minimal encoding length $I(t|s) = I(t_1|s)$ in linear time in a single right-to-left pass using the recurrence above.

This algorithm needs to be slightly modified to compute the encoding length $I_A(t)$ for scheme 1. First, the source s is omitted from input. Second, the term $l(i)$, where $1 \leq i \leq n$, now denotes the length of the longest word that starts at the ith position in target t and that also occurs at a position $j < i$ in t itself.

The algorithm can also be modified to compute the encoding length $I_A(t|s)$ for scheme 3. In this case s is not a complete sequence but a list of words. The term $l(i)$, where $1 \leq i \leq n$, now denotes the longest of the following two:

1. length of the longest word that starts at the ith position in target t and that also occurs at a position $j < i$ in t, or
2. length of the longest word that starts at the ith position in target t and that can be encoded by a pointer to list s.

It should be pointed out that all the algorithms described here run in linear time. In the case of the last algorithm, linear time is achieved under the assumption that the length of words in s is bounded by a constant.

1.3   Algorithmic Significance

In this section I review the theorems that form the basis of the algorithmic significance method. I start with the standard hypothesis testing framework where a null hypothesis and an alternative hypothesis are defined as probability distributions. Given a particular outcome of an experiment, a likelihood ratio is the probability of the outcome by the alternative hypothesis divided by the probability of the same outcome by the null hypothesis. I first show that high likelihood ratios are unlikely by the null hypothesis (lemma 1.1). I then express the alternative hypothesis in terms of encoding schemes and show that short encoding lengths are unlikely (theorem 1.1, corollary 1.1). Finally, I apply the universal null hypothesis to show that high algorithmic mutual information (the difference between the sum of individual encoding lengths and the joint encoding length) is also unlikely (theorem 1.2).

Let P_0 be a null hypothesis and P_A be an alternative hypothesis. The probabilities $p_0(t)$ and $p_A(t)$ denote the probabilities assigned to a sequence t by P_0 and P_A, respectively. The likelihood ratio for sequence t is $[p_A(t)/p_0(t)]$. Assuming $p_0(t) > 0$ for all t, the expected value $E_0[p_A(t)/p_0(t)]$ by the null hypothesis P_0 equals 1. By Markov inequality,

$$P_0\{\frac{p_A(t)}{p_0(t)} \geq c\} \leq \frac{1}{c}.$$

After taking logarithms, we obtain the following:

Lemma 1.1 *For any null hypothesis P_0 such that $p_0(t) > 0$, for every t, and for every alternative hypothesis P_A,*

$$P_0\{\log \frac{p_A(t)}{p_0(t)} \geq d\} \leq 2^{-d}.$$

This lemma may be informally summarized by stating that high likelihood ratios are unlikely by the null hypothesis. Note that in the derivation we have applied the null hypothesis twice: first to compute the probability, and then to compute the probability of a ratio of probabilities. In a typical biological application, two hypotheses are simply compared based on a commonly agreed upon standard threshold for the logarithm of the ratio, say, 3 for decimal logarithms (giving significance 10^{-3}).

Lemma 1.1 has recently been applied to determine significance of patterns detected by hidden Markov models (Barrett et al., 1997). The probability of sequence t according to model A is $p_A(t)$. The choice of hypothesis

P_0 is very flexible. For example, if we assume that every letter is generated independently with probability p_x, where $x \in \{A, G, C, T\}$ denotes the letter, then the probability of a target sequence t according to the null hypothesis is $p_0(t) = \prod_x p_x^{n_x(t)}$, where $n_x(t)$ is the number of occurrences of letter x in t. Lemma 1.1 states that, by null hypothesis, the probability that the logarithmic likelihood ratio $\log[p_A(t)/p_0(t)]$ reaches d is at most 2^{-d}. Experiments with hidden Markov models (Barrett et al., 1997) show an improved detection of amino acid sequences when the probabilities p_x are derived from the model itself.

While lemma 1.1 by itself has a potentially wide applicability, note that the term $\log[p_A(t)/p_0(t)]$ may be rewritten as $-\log p_0(t) - \log p_A(t)$. Recall that an encoding scheme that is optimal for distribution P encodes t in $-\log p(t)$ bits (e.g., Cover and Thomas, 1991). Thus, the logarithmic likelihood ratio can be interpreted as a difference in encoding lengths. I pursue this direction further by first expressing the alternative hypothesis (theorem 1.1, corollary 1.1) and then also the null hypothesis (theorem 1.2) via encoding schemes.

I begin with the alternative hypothesis P_A. Let A denote a decoding algorithm that can reconstruct the target t based on its encoding relative to the source s. $I_A(t|s)$ denotes the length of the encoding.

Make the standard assumption that A is uniquely decodable—this simply means that for every input the decoding algorithm provides at most one output (Cover and Thomas, 1991).

By the Kraft-McMillan inequality,

$$\sum_t 2^{-I_A(t|s)} \leq 1.$$

Thus, there is a normalizing constant $b \geq 1$ such that

$$\sum_t b\, 2^{-I_A(t|s)} = 1,$$

and now the alternative hypothesis can be defined as the distribution P_A that assigns probability

$$p_A(t|s) = b\, 2^{-I_A(t|s)}$$

to target t. By substituting $b\, 2^{-I_A(t|s)}$ for $p_A(t)$ in lemma 1.1, we obtain

$$P_0\{-\log p_0(t) - I_A(t|s) + \log b \geq d\} \leq 2^{-d}.$$

Finally, since $\log b \geq 0$, we obtain the following:

Theorem 1.1 *For any distribution of probabilities P_0, decoding algorithm A, and source s,*

$$P_0\{-\log p_0(t) - I_A(t|s) \geq d\} \leq 2^{-d}.$$

This theorem may be informally rephrased by stating that, by the null hypothesis, a target sequence is unlikely to have a relative encoding much shorter than the encoding that is optimal by the null hypothesis. Similar theorems have been proven in the context of competitive encoding (Cover and Thomas, 1991) and testing theory (Li and Vitányi, 1993).

Note that the source sequence s is dispensable. More precisely, theorem 1.1 may be rewritten as follows:

> **Corollary 1.1** *For any distribution of probabilities P_0 and decoding algorithm A,*
>
> $$P_0\{-\log p_0(t) - I_A(t) \geq d\} \leq 2^{-d}.$$

The direct encoding of t enables discovery of patterns in t itself. This theorem is employed below in conjunction with scheme 1 as part of application 1.

Up to this point the null hypothesis has been expressed by a probability distribution. In the following, I express the null hypothesis via an encoding scheme.

Consider the shortest encoding and the corresponding null hypothesis that is most difficult to refute. Invariance theorem (for a review, see Li and Vitányi, 1993) states that there exists a universal encoding method that gives encodings that are as short as the encodings produced by any other effective method, up to an additive constant. The decoder for the universal method is a universal prefix-free Turing machine: the shortest encoding is the shortest program for the machine that outputs target t. Such encoding leads to a null hypothesis with a very interesting property: it cannot be refuted at an arbitrary significance level by any other computable hypothesis.

To obtain a universal probability from the universal encoding method, one needs to account for programs that do not produce output because they run forever. Let A_0 denote a universal machine and let $|p|$ denote the length of a program p. The halting probability Ω (for a detailed study of Ω, see Chaitin, 1987) is the probability that A_0 halts for a p that is constructed bit by bit by random flips of a coin. That is,

$$\Omega = \sum_{p:\ A_0\ halts\ on\ p} 2^{-|p|}.$$

The probability $p_{A_0}(t)$ that a halting program outputs t is computed as

$$p_{A_0}(t) = \frac{1}{\Omega} \sum_{p:\ A_0(p)=t} 2^{-|p|}.$$

The probability distribution P_{A_0}, discovered by Solomonoff (for a review of history, see Li and Vitányi, 1993), has a remarkable property: it cannot be refuted at an arbitrary significance level by any other computable distribution. In other words, there is a constant C such that $\log[p_{A_0}(t)/p(t)]$ $\geq C$ for every computable distribution $p(t)$.

Now assume that $P_0 = P_{A_0}$. The universal coding theorem (Li and Vitányi, 1993) states that $-\log p_{A_0}(t) = I_{A_0}(t) + O(1)$, where $I_{A_0}(t)$ denotes the length of the shortest program for A_0 that outputs t. By substituting $I_{A_0}(t)$ for $-\log p_0(t)$ in theorem 1.1, and by moving the additive constant $O(1)$ into the exponent on the right side, we obtain

$$P_0\{I_{A_0}(t) - I_A(t|s) \geq d\} \leq 2^{-d+O(1)}.$$

Now assume that A is also a universal encoding method with the slight modification where the universal machine A has access to a source s.

Algorithmic mutual information $I(s;t)$ is defined as the difference $I_{A_0}(t)$ $- I_A(t|s)$. Note that algorithmic mutual information takes into account both the complexity of target t, measured by $I_{A_0}(t)$, and the similarity, measured by $I_A(t|s)$. In other words, any internal patterns in t diminish $I_{A_0}(t)$ and thus diminish $I(s;t)$ as well.

The inequality above can now be rewritten in the following compact form:

Theorem 1.2
$$P_0\{I(s;t) \geq d\} \leq 2^{-d+O(1)}.$$

Theorem 1.2 is the basis for the *extended* algorithmic significance method, which enables discovery of similarity and other dependencies in observed data via algorithmic mutual information.

Algorithmic mutual information $I(s;t)$ can be redefined as follows to be symmetrical in s and t (e.g., Li and Vitányi, 1993):

$$I(s;t) = I(t) + I(s) - I(s,t),$$

where $I(s,t)$ denotes the joint encoding length of s and t. This alternative definition is equivalent within an additive constant to the initial definition, so theorem 1.2 still holds.

Algorithmic mutual information is an ultimate, albeit impractical (noncomputable), measure of similarity: it takes into account both similarity and information content of compared objects. To obtain a practical comparison method, we need to apply encoding schemes for which encoding lengths are easy to compute. Thus, applications 2 and 3 approximately estimate the universal encoding lengths $I_{A_0}(t)$ and $I_A(t|s)$ by applying schemes 1, 2, and 3. Since these encoding schemes capture the kinds of patterns that indeed occur in the data, they may be thought of as adequate heuristic approximations.

Table 1.1 Comparative performance of algorithms for finding simple repetitive regions.

Position within HUMTPA	Repeated units	$w = 128$ $o = 64$ $d = 22$	$w = 32$ $o = 24$ $d = 20$	Guan and Uberbacher (1996)	Benson and Waterman (1994)
1017–1048	A		+	+	+
7169–7296	GT, AT	+	+	+	+
10497–10528	T		+	+	+
16897–17024	AC,TC	+	+	+	+
17089–17216	A, AAG	+	+	+	+
19153–19192	A		+	+	+
21241–21272	A, GAAAA		+		
21561–21592	TAA, TAAA		+	+	+
23888–24458	TGATAGA	+	+	+	+
26457–26496	A		+	+	+
29073–29112	A		+	+	+

1.4   Applications

I now briefly review four applications of the algorithmic significance method. Applications 1, 2, and 3 are based on schemes 1, 2, and 3, respectively; application 4 is conceptual and involves universal encoding.

1.4.1   Application 1: Discovering Simple DNA Sequences

Many genomic DNA sequences contain simple repetitive patterns. Detection of such patterns is an important component of automated DNA sequence analysis. The complete DNA sequence of the Human Tissue Plasminogen Activator gene (HUMTPA; Friezner-Degen et al., 1986; GenBank accession no. K03021) containing 36,594 bases was searched for simple repetitive patterns by applying the minimal length encoding algorithm for scheme 1. Significance was determined by corollary 1.1, assuming that, by the null hypothesis, the letters are generated independently by a uniform distribution.

The sequence was considered one window at a time, with a fixed windows length w and fixed overlap o between adjacent windows, and a significance threshold of d bits. In the original experiment (Milosavljević and Jurka, 1993a), window length w was 128 bases, the overlap o was 64 bases, and an encoding length threshold d of $22 \geq 7 + \log 36594$ bits was chosen so that the probability of significance of *any* window would be guaranteed not to exceed 0.01. This parameter setting led to the discovery of only four simple segments, as indicated in the third column of table 1.1. (The first

Human Tissue Plasminogen Activator Gene, positions 24321..24448

```
T-G-A-T-A-G-G-T-G-A-T-A-G-A-T-A-G-A-T-TGATAGAT-G-A-T-A-G-A-AGATTGATAGA
TGATAGA-T-A-C-ATAGGTGATAG-T-A-G-A-T-G-T-A-A-G-A-TGATAGATGATAGATA-GATAG
ATGATAGA-C-AGATTGATAGATGATAGA-G-A-G-A
```

Figure 1.1   Parsing of a repetitive DNA fragment.

column of the table contains the approximate location of the region within the HUMTPA gene. The second column contains the repeating unit.)

Other more recently proposed methods (Benson and Waterman, 1994; Guan and Uberbacher, 1996) claimed better sensitivity on the same benchmark, as indicated in the last two columns of table 1.1. However, the apparent superiority of the competing methods was mainly due to the fact that our initial parameters were not set for maximal sensitivity. The fourth column in table 1.1 contains the results for parameter settings $w = 32, o = 24$, and $d = 20$ proposed by Krnjajic (1996). The results with these tuned-up settings reveal sensitivity that surpasses that of the alternative methods.

The algorithm described above is accessible via electronic mail at censor @lpi.org. The electronic mail server reads the incoming mail messages that contain DNA sequences, identifies and analyzes simple and repetitive segments, and then sends the results back to the sender. The returned results include parsing of simple repetitive segments discovered by the algorithm; an example of a parsing is shown in figure 1.1.

Determining Significance of Scores

I should also mention a recently developed score-based method for detecting approximate tandem repeats (Rivals et al., 1997). A score d for an n-base sequence containing tandem repetitions of a 3-base word, including a total of m bases that do not fit the pattern, was defined as $d = 2n - 7m - 22$. By applying a simple encoding scheme and the algorithmic significance method, it has been shown that the significance of score d is 2^{-d}. An efficient algorithm was then applied to discover segments with significant scores. Note that the algorithmic significance method is here applied to determine significance of a particular score; the patterns are then discovered not by a minimal length encoding algorithm but by an algorithm that simply computes the scores.

1.4.2   Application 2: Discovering Similarity

A common repetitive pattern may cause two sequences to appear similar even though they are not related; for example, two DNA sequences that

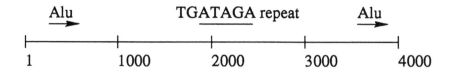

Figure 1.2   Segment 22,001–26,000 from the HUMTPA gene. (The TGATAGA repeat fragment is the same as the one represented in table 1.1 and figure 1.1.)

contain poly-A stretches may appear similar (as measured by the number of edit operations needed to transform one into the other), even though they provide little evidence about their relatedness.

To avoid the spurious similarities, "masking" procedures have been proposed (Claverie and States, 1993; Wootton and Federhen, 1993; Altschul et al., 1994). These procedures simply eliminate sequences of lower complexity from comparisons. The main drawback of these methods is that they cannot discover related sequences of lower complexity, even though the sequences themselves frequently carry enough information about their relatedness.

In the experiments described above, mutual information was computed as $I(t; s) = I(t) - I(t|s)$, where $I(t)$ is computed according to scheme 1 and $I(t|s)$ according to scheme 2. The advantage of algorithmic mutual information over other commonly used similarity measures is that it factors out the contribution of internal structure to similarity: any internal structure would lead to the decrease of $I(t)$ and, if the structure is also present in s, to the decrease of $I(t|s)$ as well; thus, any shared internal structure would not affect mutual information $I(t; s)$.

The method was applied to discover similar regions in the $22,001$–$26,000$ base pair (bp) segment of the HUMTPA gene from section 1.4.1. The segment (figure 1.2) was split into consecutive windows of length 200 with an overlap of 100 bp. Every pair of nonoverlapping windows was compared using mutual information $I(t; s)$ to identify pairs of windows that contain related sequences.

An encoding length threshold of $31 \geq 7 + 2 * \log 4000$ bits was motivated by the fact that, ignoring the additive constant in theorem 1.2, the probability that *any* pair of windows have mutual information beyond the threshold would be guaranteed not to exceed the value of 0.01. A pointer length of 6 bits was chosen for self-encoding and of 12 bits for encoding one sequence relative to the other. (It was assumed that positions of words can be encoded as the distances between consecutive common words and that the word lengths can also be differentially encoded, thus saving the total number of bits required for a pointer.)

As indicated in figure 1.2, the segment was known to contain occurrences of two Alu sequences, one between positions 253 and 545 and the other be-

Table 1.2 Three pairs of windows within the TPA segment 22,001–26,000 that exhibited highest mutual information.

Pair	$I(s;t)$	Window 1	Window 2
1	51	201–400	3601–3800
2	37	1901–2100	2201–2400
3	32	301–500	3601–3800

tween positions 3,620 and 3,911, as well as an imperfect (TGATAGA) $*$ N run between positions 1,888 and 2,458 (this region is also listed in table 1.1 and represented in part in figure 1.1). The original idea was to show that the windows containing the two occurrences of Alus would be identified while the windows containing different parts of the long TGATAGA repeat segment would not be considered similar because of their internal structure and despite their mutual similarity in terms of subword composition.

The three pairs of windows that exhibited mutual information above the threshold are shown in table 1.2. Pairs 1 and 3 correspond to the two occurrences of Alu sequences. Pair 2 consists of two windows within the repetitive pattern with an approximate TGATAGA unit. At first sight, pair 2 appeared to be a spurious similarity caused by a shared repetitive pattern. A closer inspection of the alignment of windows 1,901–2,100 and 2,201–2,400 in figure 1.3 revealed that the two windows indeed share more structure than due merely to the presence of the shared internal repeat: if we define subsequences $x = $ TGA and $y = $ TAGA, then the segment between positions 80 and 115 in window 1,901–2,100 and the segment between positions 70 and 104 in window 2,201–2,400 can both be approximately represented as

AAAyxyyxyyxTAAA.

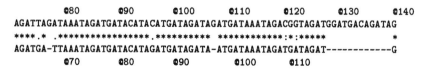

Local alignment of windows 1901-2100 (top) and 2201-2400 (bottom):

```
         ●80        ●90        ●100       ●110       ●120       ●130        ●140
AGATTAGATAAATAGATGATACATACATGATAGATAGATGATGATAAATAGACGGTAGATGGATGACAGATAG
**** .*  .***************** .********** ************:*:*****                 *
AGATGA-TTAAATAGATGATACATAGATGATAGATA-ATGATAAATAGATGATAGAT------------G
         ●70        ●80        ●90        ●100       ●110
```

Common pattern (approximate):

```
    AAA<y ><x><y ><y ><x><y ><y ><x>TAAA
```

Figure 1.3 Local alignment of windows 1,901–2,100 and 2,201–2,400 from the HUMTPA gene. The occurrence of the common pattern is indicated of the form AAA$yxyyxyyx$TAAA, where $x = $ TGA and $y = $ TAGA.

This indicates that, in addition to the simple multiplication of the TGATAGA repeat, larger units of DNA have multiplied as well, increasing the mutual information beyond the threshold.

A number of additional experiments, which I do not describe here, were performed confirming that mutual information indeed factors out similarity due to shared repetitive structures. In particular, spurious similarities that are due to poly-A segments and other simple repetitive patterns within the HUMTPA gene listed in table 1.1 did not exhibit high mutual information scores. On the other hand, sequences of high complexity and sequences that consisted of repeated modular arrangements similar to the one exhibited in our example were sensitively detected in other genes.

1.4.3   Application 3: Sequence Recognition by Hybridization

Hybridization-based techniques and others that can efficiently produce partial information about DNA fragments are emerging (Drmanac and Drmanac, 1994). The problem is to devise methods to recognize whether or not the fragment is already in the database based on the partial information produced by the high-throughput experiments.

As complete genomes of important organisms become sequenced, the emphasis of genomics may shift toward methods that provide exhaustive information about gene expression and other intracellular phenomena. Sequence recognition may become an important component of such methods; for example, sequence recognition by hybridization (Milosavljević, 1995c) has already been applied in the context of a method for studying gene expression (Milosavljević et al., 1996a).

Experiment 1 in Milosavljević et al. (1996b) demonstrates genome-scale sequence recognition by hybridization. A total of 50 *Escherichia coli* clones of 2,000 bp average length were hybridized in parallel with 997 oligomer probes. For each clone, a total of 140 words of length 7 that gave the highest hybridization signal were determined. The oligomer lists were then used for comparison against the database containing the genomic sequence of *E. coli* by employing scheme 3 and theorem 1.2. The database sequence exhibiting the highest mutual information was matched with the particular oligomer list. To determine whether the match is correct, the clone sequence was also independently determined and compared to the database sequence. A total of 33 clones (66%) were correctly recognized, indicating feasibility of genome-scale recognition.

Conditioning Factors Out Bias in Oligomer Probe Selection

One of the major obstacles preventing correct sequence recognition is the bias introduced by the initial choice of the probes for hybridization. This problem was solved by a simple conditioning of mutual information.

To demonstrate how conditioning factors out the bias, I first introduce additional notation. Let w denote the total set of words that are examined (e.g., a total of 997 such words in experiment 1 mentioned above) and let

$$I(s;t|w) = I_A(t|w) - I_A(t|s,w)$$

denote mutual information that is conditional on the choice of words w. Here $I_A(t|s,w)$ denotes the encoding length of t when the pointers can point to both s and w; note that such pointers may require more bits than the pointers that point only to s or only to w.

First consider the case where sequence t_1 happens to contain simply many words from w, even though it is not related to s. The fraction of words from t_1 in s would then typically be equal to the fraction in w. Considering that the pointers that point to both s and w require more bits than the pointers that point only to w, we may write

$$I_A(t_1|s,w) \approx I_A(t_1|w)$$

and thus

$$I(s;t_1|w) = I_A(t_1|w) - I_A(t_1|s,w) \approx I_A(t_1|w) - I_A(t_1|w) \approx 0.$$

Now consider the case when t_2 is related to s. In this case, s contains the words from w that are also in t_2. Moreover, the pointers that point to s are smaller than the pointers that point to w. Thus, we may write

$$I(s;t_2|w) = I_A(t_2|w) - I_A(t_2|s,w) \approx I_A(t_2|w) - I_A(t_2|s) > 0.$$

In contrast, the value of mutual information that is not conditional on w tends to be similar for both t_1 and t_2, because $I_A(t_1|s)$ will be similar to $I_A(t_2|s)$. Thus, conditioning factors out the bias in the choice of w and enables recognition of truly related sequences.

Direct Recognition versus Sequencing

Sequence recognition is based on a direct comparison of an oligomer list s against a target sequence t. An alternative approach is to first obtain a partial reconstruction s' of source sequence based on oligomer list s and then to compare t against s'. The reconstruction is generally referred to as *sequencing by hybridization* (Drmanac et al., 1989).

Does the intermediate reconstruction step increase specificity of recognition?

The concept of algorithmic mutual information provides a simple argument to the contrary. Assuming that mutual information is the ultimate measure of similarity, it suffices to show that

$$I(s';t) \leq I(s;t).$$

Indeed, this inequality can be proven for the generally defined algorithmic mutual information up to an additive constant:

$$I(s';t) \leq I(s;t) + O(1)$$

This algorithmic version of data processing inequality (analogous to the Shannon-entropy version; Cover and Thomas, 1991) directly follows from the fact that s' is a result of algorithmic processing of s. The inequality implies that algorithmic processing does not increase information obtained by experiments. In particular, being a result of algorithmic processing of s, the reconstructed sequence s' cannot contain more information about the underlying sequence.

1.4.4   Application 4: Discovering Global Patterns of Life

The concept of algorithmic mutual information has been employed to obtain a mathematical definition of life (Chaitin, 1979). Rather than focusing on purely biological object-level properties like reproduction or homeostasis, the definition highlights a fundamental difference in the structure of knowledge about the living and nonliving worlds. The main insight behind the definition is that in the living world patterns exist that can be observed only globally; this is in contrast to the locally observable patterns of the purely physical world.

Chaitin (1979) considers the case where a domain of observations t is split into "windows" $t_1, \ldots, t_k$ and then considers a general definition of algorithmic mutual information, which is computed as the difference between the sum of encoding lengths of individual windows $t_1, \ldots, t_k$, plus some overhead on one side, and the joint encoding length on the other side.

High mutual information occurs precisely in cases where the individual windows of observation are too small to capture a pattern: the whole pattern can be encoded more concisely by taking advantage of its regularity, which is invisible when only small pieces of the pattern are observed.

Chaitin then goes on to consider mutual information as a function of diameter D of windows: if the patterns are small, mutual information becomes negligible even for small D, while high mutual information for large D implies presence of even larger patterns that cannot be observed through windows of diameter D. Chaitin observes that the living world can be distinguished from the nonliving by the following abstract property: algorithmic mutual information in the nonliving world becomes negligible even for small diameters D while in the living world it remains high even for large D.

It follows that life can be detected at an arbitrary significance level by applying a generalized version (to allow for mutual information for more than two objects) of theorem 1.2. To detect life at significance level $2^{-d+O(1)}$, one needs to exhibit d bits of algorithmic mutual information for

```
          @1        @10      @211      @220     @271                 @292
Alu J+S   GGCCGGGCGC  ...     AGG**GAGG*  ...    AGAC*C*GTCTCAAAAAAAA
Alu  J                        TC        C        C T
Alu  S                        GC        T        T C
```

Figure 1.4   Alu subfamilies are characterized by specific bases in a number of diagnostic positions that are spread throughout their 292 bp sequence. Five diagnostic positions that contain bases specific for subfamilies Alu-J and Alu-S are indicated by asterisks. Due to large mutation rate in other nondiagnostic positions, the diagnostic positions and Alu subfamilies could not be discovered by pairwise comparisons—tens of Alu sequences had to be considered simultaneously to discover the pattern.

a sufficiently large diameter D (to remove "interference" from the patterns in the nonliving world, e.g., physical laws).

When constructing artificial examples of life to illustrate his theorem 5, Chaitin (1979) constructs hierarchical structures that resemble repetitive DNA sequences. Chaitin argues that if replication occurs at different hierarchical levels (e.g., tandem repeats of small segments vs. repetitions of larger segments that include many small segments), then the resulting pattern cannot be fully observed unless repetitions on the largest scale fit within a window. That is precisely the choice that we are implicitly making in sequence comparisons: a smaller window accommodates a single sequence with its internal repetitive patterns, while a larger window accommodates both sequences, and the problem is to decide on the size of the window. If small windows suffice for most concise encoding, then sequences are unrelated; that is, patterns are local. If larger windows give shorter encodings, then sequences are related; that is, the pattern is global.

An interesting recent example illustrating the need for large diameters of observation is the reconstruction of the evolution of Alu sequences (Jurka and Milosavljević, 1991; Milosavljević and Jurka, 1993b). The standard "bottom-up" methods for evolutionary reconstruction that are based on pairwise sequence comparisons have failed in this case: the global evolutionary pattern of Alu sequences, as illustrated in figure 1.4, remained invisible when only two Alu sequences were considered at a time (Bains, 1986). The evolutionary pattern became visible only through a "top-down" approach where a large number of sequences were considered simultaneously, as in Milosavljević and Jurka (1993b). The Alu subfamilies discovered by a "top-down" minimal length encoding approach were recently accepted by the biological community (Batzer et al., 1996).

1.5   Future Work and Generalizations

Applications of the Parsimony principle are based on a particular explicit or, more frequently, implicit measure of simplicity. Explicit measures include minimal edit distance for sequence alignment or minimal number of mutations for evolutionary reconstructions. The principle itself is sometimes questioned based on the failure of particular measures, for example, in the case of the Parsimony method for evolutionary reconstructions.

The algorithmic significance method is based on the measurement of simplicity (more precisely, complexity) in terms of encoding length (in bits). The class of patterns that can be discovered in a particular application is defined by the particular encoding scheme. A vast number of schemes may be designed. Theorem 1.1 and corollary 1.1 can be directly employed in such applications. The challenge of pattern discovery is to produce biologically meaningful schemes for which efficient algorithms exist.

Algorithmic significance decouples the problem of defining an alternative hypothesis and the null hypothesis: any null hypothesis can readily be accommodated. Recent experiments with hidden Markov models (Barrett et al., 1997) indicate that the design of the null hypothesis may be as important as the design of the alternative hypothesis (in this case, a hidden Markov model for a class of proteins). The experiments show an improved detection of amino acid sequences when the probabilities p_x are derived from the model itself. If the alternative hypothesis is a probabilistic model, then lemma 1.1 is applied (as in Barrett et al., 1997); otherwise, if the alternative hypothesis is defined via an encoding scheme, then theorem 1.1 and corollary 1.1 are applied.

Significance for particular scoring functions can be established via encoding length by applying theorem 1.1 or corollary 1.1. A scoring function for detecting possibly imperfect tandem repeats has been proposed by Rivals et al. (1997). Most important, the algorithm that detects patterns does not perform compression—it simply computes scores instead. This flexibility enables design of very efficient algorithms for pattern discovery.

Lemma 1.1 may be applied to joint distributions. In particular, P_A may be a joint distribution over (X, Y) and P_0 may be the product of marginals for X and Y. Lemma 1.1 may then be applied to discover dependencies between X and Y. It would be interesting to characterize cases where this method is more sensitive than the standard methods for detecting dependencies.

The concept of algorithmic mutual information may be used as a starting point for the design of heuristic similarity measures that take into account the complexity of compared objects. Since algorithmic mutual information is not computable and can only be approximated, theorem 1.2 is not practically useful for assigning significance values. However, it can be used as a basis for designing practical scores and also as a general heuristic principle.

An encoding scheme may be defined in the context of particular background knowledge. For example, a partially ordered set of classes of amino acids may be used as a background knowledge for defining a class of protein sequence profiles. A particular amino acid residue in a particular position may then be encoded by a path in the partially ordered set starting from the class specified in the profile for that position.

Practical pattern discovery is often interactive and involves hindsight: after observing data, a researcher may provide certain amount of additional information proving that a significant pattern is indeed present. This aspect of the discovery process can be accommodated in the encoding framework; for example, the researcher may be asked to prove significance by exhibiting a short computer program that outputs the observed data. A more practical approach would be to have some kind of controlled interactive input from researchers, allowing them in effect to function as an encoder. For example, after observing the data, researchers may make a number of choices from pulldown menus in an interactive program until they have specified the observed object; the total amount of information provided through the interface would determine encoding length and significance.

Further Information

Minimal encoding length (Kolmogorov complexity) is a measure of the information content (randomness) in the observed data. Algorithmic information theory studies this quantity in great detail (Chaitin, 1987; Cover and Thomas, 1991; Li and Vitányi, 1993) and provides a framework for a most general formulation of the principle of parsimony (Occam's Razor), originally due to Solomonoff (1964). The Parsimony principle has been widely and explicitly used in taxonomy (Sober, 1988). Applications have expanded with the appearance of macromolecular sequence data. The minimal edit distance criterion, a special case of the Parsimony principle, has been used for pairwise comparisons of macromolecular sequences (Waterman, 1984). The principle has also been applied in statistical inference (Rissanen, 1989). A perspective on machine discovery from the viewpoint of minimal length encoding has been summarized in Milosavljević (1995b).

Chapter 2

Assembling Blocks

Jorja G. Henikoff

A *block* is an ungapped local multiple alignment of amino acid sequences from a group of related proteins. Ideally, the contiguous stretch of residues represented by a block is conserved for biological function. Blocks have depth (the number of sequences) and width (the number of aligned positions). There are currently several useful programs for finding blocks in a group of related sequences that I do not discuss in detail here. Among these, Motif (Smith et al., 1990) and Asset (Neuwald and Green, 1994) both align blocks on occurrences of certain types of patterns found in the sequences; Gibbs (Lawrence et al., 1993; Neuwald et al., 1995) and MEME (Bailey and Elkan, 1994) both look for statistically optimal local alignments; and Macaw (Schuler et al., 1991) and Somap (Parry-Smith and Attwood, 1992) both give the user assistance in finding blocks interactively.

After candidate blocks are identified by a block-finding method, they can be evaluated and assembled into a set representing the protein group, resulting in a multiple alignment consisting of ungapped regions separated by unaligned regions of variable length (Posfai et al., 1988; figure 2.1). The block assembly process is the subject of this chapter. Both the Blocks (Henikoff and Henikoff, 1996a) and Prints (Attwood and Beck, 1994) databases consist of such sets of blocks and between them currently represent 1,163 different protein groups. These collections of blocks are more sensitive and efficient for classifying new sequences into known protein groups than are collections of individual sequences, as demonstrated by comprehensive evaluations (Henikoff and Henikoff, 1994b, 1997), by genomic studies (Green et al., 1993), and by individual studies (Posfai et al., 1988; Henikoff, 1992, 1993; Attwood and Findlay, 1993; Pietrokovski, 1994; Brown, 1995).

```
BAH_STRHY ( 67) RTLLYLHGGSYALGS (18) VLALHYRRPPESPFPAAVEDAVAAY ( 6) GCPPGRVTLAGDSAGAGLA (150)
EST_ACICA ( 73) QLIFHIHGGAFFLGS (18) VIHVDYPLAPEHPYPEAIDAIFDVY ( 6) GIKPKDIIISGDSCGANLA (147)
LIP2_MORSP (159) AAMLFFHGGGFCIGD (18) VVSVDYRMAPEYPAPTALKDCLAAY (10) GASPSRIVLSGDSAGGCLA (188)
YBAC_ECOLI ( 85) ATLFYLHGGGFILGN (18) VIGIDYTLSPEARFPQAIEEIVAAC (10) QINMSRIGFAGDSAGAMLA (148)
LIPS_HUMAN (344) SLIVHFHGGGFVAQT (18) IISIDYSLAPEAPFPRALEECFFAY (10) GSTGERICLAGDSAGGNLC (345)
 LIPS_RAT (343) ALVVHIHGGGFVAQT (18) IISIDYSLAPEAPFPRALEECFFAY (10) GSTGERICLAGDSAGGNLC (339)
```

Figure 2.1 A set of blocks from the Blocks database representing the lipolytic "G-D-X-G" enzyme family. There are three blocks in six sequences (BAH_STRHY, etc.). The numbers in parentheses indicate the extent of the unaligned regions outside of the blocks.

Issues

Issues that must be addressed during block assembly include the number of blocks provided to the assembly module by the block finders, block width, the number of times a block occurs in each sequence (zero to many), overlap of blocks, and the order of multiple blocks within each sequence. Once these issues are decided, it is necessary to score individual competing blocks and then competing sets of blocks. Depending on the goals of the assembly step, blocks may compete for inclusion in the final set, with some being dropped. Individual blocks compete if they overlap when nonoverlapping blocks are desired, for instance. Sets of blocks compete in number of sequences and number of blocks included. Table 2.1 summarizes how different block systems address these issues.

2.1 Algorithmic Techniques

Most block finders provide some assistance with assembling candidate blocks. Macaw and Somap facilitate manual assembly. Given information about the number, widths, and number of occurrences of blocks, MEME and Gibbs look for sets of blocks with those characteristics. In addition, Gibbs optionally requires multiple blocks to be in the same order in all sequences. The Motif and Asset programs find large numbers of blocks and then group those that are geographically close.

The Protomat Block Maker system (Henikoff and Henikoff, 1991; Henikoff et al., 1995) features a separate module, Motomat, that was specifically designed to assemble blocks automatically. The candidate blocks presented to Motomat can come from several different block finders. Our World Wide Web (WWW) implementation (http://blocks.fhcrc.org) currently assembles blocks from Motif or from Gibbs. Motomat begins with candidate blocks and the full-length sequences for all of the sequences included in the blocks. It then proceeds to find a best set of nonoverlapping blocks that occur in the same order in a minimum critical number of the sequences: it merges blocks that overlap compatibly in all sequences, resulting in wider blocks; scores each merged block; extends each merged block out to a max-

Table 2.1 Issues affecting block assembly.

Program	Program approach
	Number of blocks produced by block finder
Motif	No theoretical maximum
Asset	No theoretical maximum
MEME	User specifies
Gibbs	User specifies
Macaw	No theoretical maximum
Somap	No theoretical maximum
	Block width
Motif	Motif width
Asset	Pattern width
MEME	User specifies maximum width
Gibbs	User specifies minimum width
Macaw	User decides
Somap	User decides
	Number of occurrences of block in sequences
Motif	Specify total number in all sequences
Asset	Specify total number in all sequences
MEME	Specify total number in all sequences
Gibbs	Specify total number in all sequences
Macaw	User can leave out sequences
Somap	User can leave out sequences
	Blocks can overlap
Motif	Yes, attempts to combine
Asset	Yes, combines
MEME	No
Gibbs	No
Macaw	If user inserts gaps
Somap	If user inserts gaps
	Block order maintained in all sequences
Motif	No
Asset	No
MEME	No
Gibbs	Optional
Macaw	Yes
Somap	Yes

imum width by maximizing the block score; drops low-scoring blocks based on block score; enumerates and scores candidate sets of blocks; and finally selects a single highest scoring set.

Motomat requires four parameters: the minimum number of sequences that must be included in the final set of blocks, the minimum individual block score, the maximum final block width, and an indication of whether the blocks may appear more than once in the sequences. All parameters are set automatically by default. If the minimum number of sequences is not propagated from the block finder, it is the lesser of half the total number of sequences plus three and the total number of sequences. This formula was derived by analyzing several versions of the Blocks database. The minimum individual block score is taken to be one standard deviation below the mean block score. This value was selected based on experience after making several versions of the Blocks database. The maximum final block width is currently set to 55 for display purposes (the average width is about 30). The number of times a block may occur in each sequence is propagated from the block finder.

2.1.1   Scoring Individual Blocks

Motomat's assembly steps are straightforward, but the results depend entirely on scoring, first of individual blocks and then of sets of blocks. Throughout the scoring process it is important to emphasize certain features to obtain consistently good results. If this is not done, marginal blocks can displace strong blocks in the final result.

Scoring an individual block in isolation is problematic (see Schuler et al., 1991, on Macaw, for an excellent discussion). Perfectly reasonable-looking blocks can be produced by block finders from nonsense sequences (Henikoff, 1991). Weak blocks can achieve high scores based on the similarity of a subset of sequences or simply because no better blocks are found. The Motomat block score is based on the sum of pairs score, or *SP-score* (Smith et al., 1990; Schuler et al., 1991), for each column of the alignment, which requires a substitution matrix; Motomat's default is Blosum62 (Henikoff and Henikoff, 1992). However, rather than simply adding the SP-score for each column, the mean, called the *MP-score*, is computed, and then only the positive MP-scores are summed. This approach recognizes that, even within a conserved region, some columns are not conserved. To compare blocks with one other, we divide the sum of the positive MP-scores by the square root of the block width so that wider blocks do not dominate. Finally, Motomat computes the mean and standard deviation of all the block scores and drops blocks that score lower than one standard deviation below the mean. The objective is to eliminate weak blocks early in the assembly process in order to concentrate the search for a best set on the stronger blocks. Elimination of weak blocks can also significantly decrease the number of competing candidate sets of blocks.

Macaw addresses block scoring more elegantly by applying an extension of Karlin-Altschul statistics to assign a P value to a block score computed purely from SP-scores (Schuler et al., 1991). However, this approach is not without its difficulties because a high degree of similarity between just two of several sequences in a block can produce a high SP-score. Nevertheless, the statistical approach is attractive and ways to incorporate it into Motomat are being considered. Another modification being tested incorporates sequence weighting (Henikoff and Henikoff, 1994a) into the SP-score to prevent a few closely related sequences from dominating. Most block finders are susceptible to being misled by this situation. For example, Asset addresses it by providing a purge program to preprocess the sequences by removing similar ones (Neuwald and Green, 1994).

2.1.2   Scoring Sets of Blocks

Designing a scoring scheme for sets of blocks depends on the desired outcome. Motomat tries to assemble sets of blocks so that they appear in the same order in a critical number (at least half) of the sequences without overlapping. However, if told that the group of sequences is expected to contain repeated domains, it will drop the order requirement. In candidate sets, the same sequences occur just once in every block. This is a somewhat artificial restriction since some sequences may logically occur more than once (repeated domains) or not at all (fragments) in some blocks. It is enforced mainly to facilitate the efficient enumeration of possible sets of blocks. We are considering adding sequence occurrences to the final blocks after they have been determined.

Motomat tries to balance the number of sequences in the best set with the number of blocks; more sequences are better, but so are more high-scoring blocks. The set score is basically the sum of the participating block scores multiplied by the percentage of sequences included in the blocks. However, it is necessary to give certain blocks a bonus so that candidate sets including them have an advantage. These are the blocks formed by merging together several candidate blocks early in the assembly process, indicating that the block finders discovered them more than once. This is done simply by amplifying the block score by multiplying it by the number of merged blocks. This technique works better with block finders such as Motif that report many blocks than with those such as Gibbs that try to give more precise results and therefore tend not to produce multiple blocks in the same region.

Enumeration of candidate sets of blocks is accomplished by a depth first search through an acyclic graph that has the blocks for nodes and arcs connecting pairs of blocks that do not overlap and that have the same order in at least the critical number of sequences (Henikoff and Henikoff, 1991). The search through the graph stops when the number of sequences in all of the blocks visited so far falls below the critical number. A similar but more rigorous algorithm for block assembly is described by Z. Zhang et al. (1994). Each candidate set is then scored and the highest scoring set selected.

2.2   Future Work and Generalizations

The Protomat Block Searcher system was designed to run automatically to process large numbers of protein families without requiring human inter-

```
PPAR_MOUSE 102 CRICGDKASGYHYGVHACEGCKGFFRRTIRLKLVYDK
7UP1_DROME 200 CVVCGDKSSGKHYGQFTCEGCKSFFKRSVRRNLTYSC
E75A_DROME 245 CRVCGDKASGFHYGVHSCEGCKGFFRRSIQQKIQYRP
THA1_MOUSE  53 CVVCGCKATGYHYRCITCEGCKGFFRRTIQKNLHPTY

PPAR_MOUSE 135                      VYDKCDRSCKIQKKNRNKCQYCRFHKCLSVGM
7UP1_DROME 234                      YSCRGSRNCPIDQHHRNQCQYCRLKKCLKMGM
E75A_DROME 280                      RPCTKNQQCSILRINRNRCQYCRLKKCIAVGM
THA1_MOUSE  89                      YSCKYDSCCVIDKITRNHCQLCRFKKCIAVGM
```

Figure 2.2   Overlapping blocks during block assembly. These two candidate blocks cannot be merged because they overlap differently in different sequences. They will compete for inclusion in the best set. The overlapping amino acids appear in boldface. Sequence names and offsets appear on the left.

vention. In addition to being used to make the Blocks database, the system processes about 25 requests daily on our e-mail and WWW servers. In its attempt to determine good parameter settings for the sequences presented to it, it sometimes produces suboptimal results. Comparison of the Blocks database, made automatically by Protomat, with the Prints database reveals both the power and the shortcomings of Protomat. The power is that, depending on the block finder used, it can process hundreds of protein groups in a few days whereas 50 new groups can be added to Prints every three months only with great effort. However, the Prints blocks are more consistently correct. In particular, manual systems like Somap and Macaw allow the insertion of a few gaps to correct alignments. Blocks that overlap differently in different sequences can be manually separated using these programs, whereas Motomat is likely to discard one (figure 2.2). Although ways to introduce very limited gaps into candidate blocks have been considered to correct obvious misalignments before Motomat's block merging step, retaining the ungapped definition of a block by dividing it, as is currently done manually for the Prints database, is preferred.

In addition to trying to assess the statistical significance of an individual block, ways to use the ability of a position-specific scoring matrix (PSSM; Henikoff and Henikoff, 1994b) computed from a block to separate true positive and true negative sequences in a database search are being researched as an aid in scoring a block. Currently a PSSM computed from each block in the Blocks database is searched against the same protein sequence database from which the block is derived to assess this ability, which is called "strength" (Henikoff and Henikoff, 1991). This number can be estimated theoretically for integer PSSMs (Tatusov et al., 1994), and this technique is used to calibrate blocks from the Prints database. It may be feasible for Motomat to estimate the strength of candidate blocks to help decide whether to drop them. This method is attractive because it simulates the major use of blocks, which is to classify sequences of unknown function into known protein families.

Chapter 3

MEME, MAST, and Meta-MEME: New Tools for Motif Discovery in Protein Sequences

Timothy L. Bailey, Michael E. Baker,
Charles P. Elkan, and William N. Grundy

3.1 Software for Finding Sequence Motifs

We are in the midst of an explosive increase in the number of DNA and protein sequences available for study, as various genome projects come on line. This wealth of information offers important opportunities for understanding many biological processes and developing new plant and animal models, and ultimately drugs, for human diseases, in addition to other applications of modern biotechnology. Unfortunately, sequences are accumulating at a pace that strains present methods for extracting significant biological information from them. A consequence of this explosion in the sequence databases is that there is much interest and effort in developing tools that can efficiently and automatically extract the relevant biological information in sequence data and make it available for use in biology and medicine.

In this chapter, we describe one such method that we have developed based on algorithms from artificial intelligence research. We call this software tool MEME (Multiple Expectation-maximization for Motif Elicitation). It has the attractive property that it is an "unsupervised" discovery tool: it can identify motifs, such as regulatory sites in DNA and functional domains in proteins, from large or small groups of unaligned sequences. As we show below, motifs are a rich source of information about a dataset;

they can be used to discover other homologs in a database, to identify protein subsets that contain one or more motifs, and to provide information for mutagenesis studies to elucidate structure and function in the protein family as well as its evolution.

3.1.1 MEME—An Unsupervised Learning Tool

Learning tools are used to extract higher level biological patterns from lower level DNA and protein sequence data. In contrast, search tools such as BLAST (Basic Local Alignment Search Tool) take a given higher level pattern and find all items in a database that possess the pattern. Searching for items that have a certain pattern is a problem intrinsically easier than discovering what the pattern is from items that possess it. The patterns considered here are motifs, which for DNA data can be subsequences that interact with transcription factors, polymerases, and other proteins. For proteins a motif may be a subsequence that binds to DNA, to other proteins, to regulatory ligands (e.g., steroids), or to substrates.

Learning tools can be divided into supervised and unsupervised tools. A supervised learning tool takes as input a set of sequences and discovers a pattern that each one of the sequences shares. The word "supervised" here refers to the fact that a human acting as a teacher or supervisor must screen the input set of sequences carefully before giving the set as input to the learning tool. Supervised learning is often done by humans rather than by software because it is an open-ended problem that is harder than searching, as just mentioned. Examples of human supervised learning are creating profiles (Gribskov et al., 1987) and PROSITE signatures (Bairoch, 1992). Both require multiple alignment and extensive screening of the sequences by humans, in contrast to MEME, as we discuss below.

An unsupervised learning tool takes as input a set of sequences and discovers a pattern that some of the sequences share. Relaxing the requirement that all sequences contain a pattern allows analysis of less carefully sorted input data. As a result, there is no need for a human supervisor to screen each member of the set of sequences with great care. Reducing the amount of human preprocessing of data allows more data to be analyzed in less time, an important advantage for genome projects. Moreover, motifs of biological importance can be identified that may escape analysis by humans using traditional approach.

However, it is a more complicated task to develop tools for unsupervised learning than for supervised learning, because the space of possible patterns in unsupervised learning is much larger. The pattern to be discovered is not required to be in any given input sequence, yet it may occur repeatedly in a single sequence. The unsupervised learning algorithm must simultaneously look for a cluster of input sequences and a pattern that the members of this cluster have in common. MEME performs unsupervised learning.

An important advantage of MEME is that the motif analysis is not prevented by the inclusion of nonhomologous proteins in the training set. MEME can identify these proteins and exclude them from the analysis. This property can be especially important in analyses of small datasets, which is often all that biologists have. Sometimes a dataset includes a sequence with a BLAST score that is borderline for homology with the rest of the set. The temptation is to include the borderline sequence in the training set, if the set is small and one needs as many sequences as possible for analysis. The information from a distantly related protein can improve the sensitivity of motifs for database searches as well as provide important information about conserved structures.

With the above considerations in mind, this chapter investigates the effectiveness of MEME and MAST (Motif Alignment Search Tool) with a small training set of 10 diverse dehydrogenases that include 11β-hydroxysteroid dehydrogenase (11β-HSD) types 1 and 2, and four serine proteases that belong to a different protein superfamily. Using this training set tests how effectively MEME analyzes dehydrogenases in the presence of four proteins that are not related to dehydrogenases. This analysis also shows whether MEME can find important motifs in the four serine proteases in the presence of 10 dehydrogenases. Our results indicate that MEME is highly useful for small training sets, which complements our previous reports that MEME can identify motifs in a larger dataset of 37 short-chain alcohol dehydrogenases (Bailey et al., 1997; Grundy et al., 1997) and that MAST can use these motifs to search Genpept and identify distant homologs with almost perfect sensitivity and specificity.

While using MEME we have discovered that its selectivity for identifying motifs from distinct protein families makes it an excellent tool for multiple local alignments (Grundy and Elkan, 1997). Because they effectively ignore nonhomologous proteins, MEME motifs give rise to a high-quality multiple alignment. In contrast, the most widely used methods (Feng and Doolittle, 1987; Higgins et al., 1996) use a progressive alignment approach, which will fit a nonhomologous protein to the rest of the dataset, leading to an inaccurate alignment.

3.1.2   Hydroxysteroid Dehydrogenases and Their Homologs

The short-chain alcohol dehydrogenase (Baker, 1991; Persson et al., 1991; Tannin et al., 1991; Krozowski, 1992; Jornvall et al., 1995) protein family, also called the sec-alcohol dehydrogenase family (Baker, 1994b), contains mammalian 11β-HSD and 17β-HSD, which are vital in regulating the concentrations of androgens, estrogens, and adrenal steroids in humans (figure 3.1). The importance of steroids in human physiology and in the growth of breast and prostate tumors has been a major stimulus for research to un-

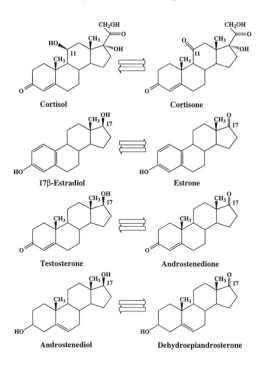

Figure 3.1 Reactions catalyzed by 11β-HSD and 17β-HSD. In humans, 11β-HSD type 1 catalyzes the conversion of the inactive steroid cortisone to cortisol, the biologically active glucocorticoid. The type 2 enzyme catalyzes the reverse direction. Similarly, various 17β-HSDs catalyze either oxidation or reduction of estrogens and androgens. 17β-HSD type 1 catalyzes reduction of estrone to estradiol, type 2 catalyzes oxidation of estradiol to estrone, type 3 catalyzes reduction of androstenedione to testosterone, and type 4 catalyzes the oxidation of estradiol to estrone.

derstand the function of steroid dehydrogenases and their homologs. This is a large enzyme family that is functionally and phylogenetically diverse, with examples in bacteria, plants, and animals. As expected from such diversity, many pairwise sequence comparisons reveal less than 25% identity after adding gaps to the alignment (Persson et al., 1991; Baker, 1994c). Thus, we have a large set of distantly related enzymes that can test MEME's efficiency and accuracy in identifying motifs characteristic of all or a subset of the enzyme family, and for searching the database with these motifs to identify more distantly related homologs, an important application of motif analysis.

Moreover, the 3D structures of four members of the family—*Streptomyces hydrogenans* 20β-HSD (Ghosh et al., 1991, 1994a), rat dihydropteridine reductase (Varughese et al., 1992, 1994), human 17β-HSD type 1 (Ghosh et al., 1994b), and plant enoyl-acyl-carrier protein reductase (Rafferty et al., 1995)—have been determined. Thus, we can test the hypothesis that motifs reported by MEME correlate with secondary and tertiary structure.

3.1.3   Serine Proteases

Like the steroid dehydrogenases and their homologs, the serine protease family is phylogenetically diverse and important in many physiological processes from metabolism of proteins, to tumor metastasis, to embryonic and fetal development. The serine proteases were used by Lipman et al. (1989) as a development dataset for their multiple alignment program. One reason for choosing the serine proteases is that their 3D structures are known (Greer, 1991), so Lipman et al. could evaluate the correctness of alignments with respect to superimposed 3D structures.

Here we show that MEME analysis of a dataset of 10 dehydrogenases and four serine proteases can identify motifs characteristic of each protein family. Moreover, we port MEME motifs into two searching tools that we have developed, MAST and Meta-MEME (Bailey and Elkan, 1994, 1995; Bailey et al., 1997; Grundy et al., 1997) and extract homologs for each protein family among the 200,000 sequences in Genpept98.

3.2   Methods

The MEME software tool (Bailey and Elkan, 1994, 1995) takes a group of related amino acid or nucleic acid sequences and produces a set of motifs, which describe the group. The software includes programs for annotating sequences with the motifs and displaying schematic diagrams of sequences in terms of the motifs.

Typically, MEME is used to discover a set of motifs that describe a group of related sequences. Chosen motifs of interest can be used to search a sequence database, such as SWISS-PROT (Bairoch, 1994) or Genpept, in a detailed manner using MAST. Such a search allows one to detect new members of the group and to evaluate the sensitivity and selectivity of the individual motifs. The MAST program prints the sequences from the database that contain matches to one or more of the motifs and indicates where the matches occur. MAST also prints schematic diagrams of each sequence with one or more matches to motifs and orders the sequences according to a scoring function that considers all the motifs.

The MEME system is written in the C programming language and currently runs on Sun SparcStation and DEC Alpha computers under the UNIX operating system. A parallel version of MEME runs on the Cray T3E scalable parallel computer installed at the San Diego Supercomputer Center. This version of MEME is publicly available through a World Wide Web server at http://www.sdsc.edu/MEME.

3.2.1   Discovering Motifs

A motif is a pattern that describes subsequences of fixed length N and that do not contain gaps. In this chapter, a motif is a probability model that assigns a certain probability to each subsequence of length N. If a subsequence matches the model, it is called an *instance* of the motif. The degree to which a subsequence is an instance of a motif is the probability of the subsequence according to the motif model.

The probability model representing a motif is a matrix of probabilities. The number of columns of the matrix representing a motif is the length of the motif, N, that is, the length of subsequences that match the motif. The number of rows is the size of the alphabet: four for DNA motifs and 20 for protein motifs. The entry of the matrix at row number i and column number j is the probability of finding the letter numbered i at the jth position in a subsequence that is an instance of the motif. For proteins, the so-called letters are amino acids, which are also known as residues. For DNA sequences, each letter is a nucleic acid, that is, A, C, G, or T. The motif matrix can be thought of as the frequencies that would be observed in each column of a multiple alignment of all possible examples of the pattern.

MEME discovers multiple motifs by using a statistical algorithm called expectation-maximization (EM; Dempster et al., 1977) to fit a series of motif models to a group of related sequences. When MEME discovers multiple motifs, it produces one probability matrix per motif discovered. The fact that MEME discovers multiple motifs means that patterns containing gaps can still be discovered; MEME will split them into multiple, ungapped motifs.

MEME automatically determines the width of each motif and decides if each pattern occurs in all or only a subset of the sequences. If the sequences are believed by the user to contain repeats of a single pattern, MEME can use this information by changing its statistical model under the control of the user. The default is to assume that each pattern occurs no more than once in each sequence in the dataset. This default assumption encodes some background knowledge that is typically true about DNA motifs and motifs in globular proteins.

The MEME algorithm finds motifs one at a time in the group of sequences under analysis. It searches for motifs that describe statistically significant patterns in the sequences. For each motif, MEME maximizes a heuristic function (Bailey and Elkan, 1995), which balances the width, crispness, and coverage, that is, the number of sequences in the dataset that contain matches to the motif. For a given width, a motif that closely matches positions in each of the sequences is more statistically significant than one that matches fewer sequences equally well. This trade-off is made by MEME by choosing the motif of a given width with the maximum likelihood.

Comparing motifs of different widths is more problematic. MEME uses a heuristic function that combines the likelihoods and widths of motifs into one statistic. This heuristic function is based on a standard likelihood ratio test. In analyzing a series of different motif widths, MEME chooses the width that maximizes the statistical heuristic function. For each width that it tries, MEME repeatedly executes the EM algorithm from different starting points and chooses the final motif of the given width that maximizes the likelihood function. In this way, MEME ensures that the motifs found are likely to be the most statistically significant ones present in the group of sequences.

The unsupervised nature of MEME eliminates unconscious human bias in locating the boundaries of a motif. Once an optimum motif has been found, MEME continues examining the dataset for more motifs up to the number requested by the user. MEME avoids finding overlapping motifs by storing the motifs it has found so far and combining this information with the prior probabilistic model of the next motif to be found.

For each motif, MEME reports

- a matrix giving the probability of each residue at each position in the motif,
- the most likely location of the motif in each sequence in the dataset,
- a plot of the information content at each position of the motif,
- a consensus sequence summary of the motif, and
- a position-dependent scoring matrix.

Unlike the probability matrix, which gives the expected frequency of each amino acid at each position of the motif, the scoring matrix is a log-odds matrix that also takes into account the background probability of each amino acid appearing outside the motif. The scoring matrix is used in searching for matches to the motif.

3.2.2   Searching with Motifs

The MEME system incorporates the position-dependent scoring matrix into two tools, MAST and Meta-MEME, that are used for homology searches and to visualize the structure of a family of sequences. These tools are described briefly here.

The MAST Tool

MAST searches a database for matches to a set of motifs and generates a histogram of match scores for each motif. The tool also shows where and how each motif matches each sequence in a database and provides a block diagram in terms of motifs of the sequences with one or more matches. The search is conducted using the position-dependent scoring matrix for each motif produced by MEME. For each motif, the subsequence beginning at each position in every sequence under investigation is scored using the motif's scoring matrix. All sequence positions scoring above a threshold computed by MEME or provided by the user are printed. The score of a position in a sequence gives a measure of the degree of match of the motif to that position.

Matches are determined by scoring the sequence using the MEME-generated scoring matrices and thresholding with either the MEME-generated or a user-provided threshold. The standard output of MAST is to show each sequence with one or more matches in its entirety, with all matches to all motifs indicated. In addition, this information is abstracted into a block diagram where just the motif matches and their spacing are shown for each sequence. Since the order and spacing of motif matches are important, especially for distant homologs, the block diagrams make it easier for the user to discriminate real homologs from chance matches.

MAST also computes a score for each sequence that sums the highest match score for each motif. This max-sum score is used to sort the block diagrams for the sequences that meet the criterion of at least one motif match above the threshold. Since it combines measures of similarity between the sequence and several motifs, the max-sum score tends to detect distant similarities more selectively than individual motif scores.

For example, as discussed further below, MEME finds seven motifs specific for steroid dehydrogenases and their homologs (figure 3.2), which can be used for MAST analysis of the training set (figure 3.3). The motif diagram in figure 3.3 shows that 20β-HSD has high scores for all seven motifs, which is reflected in the high max-sum score and an E value of 2.3×10^{-62}. In contrast, 11β-HSD type 1 has high scores for only motifs 1, 2, and 7. Although the other motifs are found in this enzyme (Bailey et al., 1997), their scores are below the cutoff for inclusion in the motif diagrams. (One can do an analysis with a lower threshold to verify that the motifs are weakly present.) Nevertheless, the "max-sum" is high enough to yield an E value of 2.7×10^{-23}, which conclusively shows the homology of 20β-HSD and 11β-HSD type 1.

Moreover, the order of motifs in both proteins is 1, followed by 7, followed by 2, and motifs 7 and 2 are nine amino acids from each other in both 20β-HSD and 11β-HSD type 1. Motifs 1 and 7 are 110 amino acids and 107 motifs from each other in 11β-HSD type 1 and 20β-HSD, respectively. The conservation of motif order and spacing is consistent with a common 3D structure for these homologous proteins.

Contrast the above with bovine chymotrypsin, which has a low score for motif 1, which is positioned 51 amino acids from the carboxy-terminus. Similarly motif 2 in pig elastase is only 19 amino acids from the amino-terminus instead of being close to the carboxy-terminus. Thus, we can consider the scores for chymotrypsin and elastase as random similarities for small segments.

Meta-MEME

Recently, hidden Markov models (HMMs) have been applied to the task of modeling families of related proteins. These models are attractive be-

```
MOTIF  1                              bits 4.9
                                           4.2
                         Information  3.6
                         content      3.0 *
                         (41.5 bits)  2.4 *     **    * *
                                      1.8 *    ***   *** * *
                                      1.2 ******** **** * **
                                      0.6 ******************
                                      0.0 ------------------
                         Multilevel       KVALVTGAASGIGLExAK
                         consensus        VII  G R   RA  R
Sequence name            Start  Score         Site
11beta-HSD-TYPE1          35    45.20     KKVIVTGASKGIGREMAY
11beta-HSD-TYPE2          82    37.10     RAVLITGCDSGFGKETAK
15-OH Prostaglandin DH    6     49.56     KVALVTGAAQGIGRAFAE
20beta-HSD Streptomyces   7     43.12     KTVIITGGARGLGAEAAR
3beta-HSD Bacterial       7     46.24     KVALVTGGASGVGLEVVK
ActIII protein  Streptomyces 7  46.03     EVALVTGATSGIGLEIAR
7alpha-HSD Eubacterium    7     47.00     KVTIITGGTRGIGFAAAK
ADH Drosophila            8     27.46     KNVIFVAGLGGIGLDTSK
Ribitol DH Klebsiella     15    51.35     KVAAITGAASGIGLECAR
3-Oxoacyl-[Acyl-Carrier Prot] 6 46.60     KIALVTGASRGIGRAIAE
```

```
MOTIF 2                               bits 4.9
                                           4.2
                         Information  3.6        *
                         content      3.0 *    *       *
                         (28.5 bits)  2.4 *    *       *
                                      1.8 * ***   * **
                                      1.2 ******* ***
                                      0.6 ***********
                                      0.0 ------------
                         Multilevel       YSASKAAVVGFT
                         consensus        G
Sequence name            Start  Score         Site
11beta-HSD-TYPE1          183    30.65     YSASKFALDGFF
11beta-HSD-TYPE2          231    17.10     YGTSKAAVALLM
15-OH Prostaglandin DH    151    35.16     YCASKHGIVGFT
20beta-HSD Streptomyces   152    27.86     YGASKWGVRGLS
3beta-HSD Bacterial       150    29.87     YSASKAAVSALT
ActIII protein Streptomyces 157  39.20     YSASKHGVVGFT
7alpha-HSD Eubacterium    157    29.42     YPASKASVIGLT
ADH Drosophila            153    30.37     YSGTKAAVVNFT
Ribitol DH Klebsiella     158    29.44     YTASKFAVQAFV
3-Oxoacyl-[Acyl-Carrier Prot] 151 25.45    YAAAKAGLIGFS
```

Figure 3.2 MEME motifs. (a) The first 12 motifs from MEME analysis of short-chain alcohol dehydrogenases. The information content plot is a measure of conservation at each position of the motif. The consensus sequence below each information content plot shows sites where specific amino acids are present with a probability of at least 20%. Motifs 1–3 and 6–9 are characteristic of dehydrogenases; motifs 4, 5, and 10–12 are characteristic of serine proteases.

MOTIF 3

```
                           bits 6.1
                                5.5
                                4.9
                                4.2
            Information        3.6
            content            3.0            *      **
            (37.7 bits)        2.4      *     *     ** *
                                1.8      **  **  ******
                                1.2      **  *********
                                0.6  **************
                                0.0  ---------------
            Multilevel              ExFGRLDVLVNNAGI
            consensus                 Y    V I
```

Sequence name	Start	Score	Site
11beta-HSD-TYPE2	155	9.22	TTSTGLWGLVNNAGH
15-OH Prostaglandin DH	80	43.96	DHFGRLDILVNNAGV
20beta-HSD Streptomyces	76	42.81	EEFGSVDGLVNNAGI
3beta-HSD Bacterial	75	38.65	RRLGTLNVLVNNAGI
ActIII protein Streptomyces	79	41.48	ERYGPVDVLVNNAGR
7alpha-HSD Eubacterium	81	43.34	QKYGRLDVMINNAGI
ADH Drosophila	82	31.36	AQLKTVDVLINGAGI
Ribitol DH Klebsiella	84	25.99	QLTGRLDIFHANAGA
3-Oxoacyl-[Acyl-Carrier Prot]	75	45.42	AEFGEVDILVNNAGI

MOTIF 4

```
                           bits 6.1
                                5.5
                                4.9 *           *        *
                                4.2 *           *       **
            Information        3.6 *           *       **
            content            3.0 *           *       **
            (58.5 bits)        2.4 **    *     *       **
                                1.8 ******     ** *****
                                1.2 ******* *********
                                0.6 ****************
                                0.0 ----------------
            Multilevel              CGGSLINENWVVTAAHC
            consensus                       L
```

Sequence name	Start	Score	Site
TRYPSIN BOVINE	31	60.64	CGGSLINSQWVVSAAHC
CHYMOTRYPSIN BOVINE	42	65.08	CGGSLINENWVVTAAHC
ELASTASE PIG	56	60.70	CGGTLIRQNWVMTAAHC
THROMBIN BOVINE	394	55.74	CGASLISDRWVLTAAHC

Figure 3.2 (a) continued.

MOTIF 5

```
                            bits 6.1
                                 5.5
                                 4.9 *            *
                                 4.2 *            *
          Information            3.6 *            *
          content               3.0 *   *    *   *
          (38.4 bits)           2.4 **** ***   *
                                 1.8 *******   *
                                 1.2 **********
                                 0.6 **********
                                 0.0.----------
          Multilevel                CQGDSGGPLVC
          consensus
```

Sequence name	Start	Score	Site
TRYPSIN BOVINE	179	42.66	CQGDSGGPVVC
CHYMOTRYPSIN BOVINE	191	40.73	CMGDSGGPLVC
ELASTASE PIG	210	41.66	CQGDSGGPLHC
THROMBIN BOVINE	567	32.72	CEGDSGGPFVM

MOTIF 6

```
                            bits 6.1
                                 5.5
                                 4.9
                                 4.2                *
          Information            3.6                *
          content               3.0      *          *
          (37.5 bits)           2.4 * *    **    * *
                                 1.8 **** * **    ***
                                 1.2 ****** **** ***
                                 0.6 ***************
                                 0.0 ---------------

          Multilevel                IRVNAVAPGFVxTPM
          consensus
```

Sequence name	Start	Score	Site
11beta-HSD-TYPE1	165	14.28	IVVVSSLAGKVAYPM
15-OH Prostaglandin DH	176	36.50	VRLNAICPGFVNTAI
20beta-HSD Streptomyces	175	46.29	IRVNSVHPGMTYTPM
3beta-HSD Bacterial	176	35.55	RRVNSIHPDGIYTPM
ActIII protein Streptomyces	180	49.68	ITVNAVCPGFVETPM
7alpha-HSD Eubacterium	180	38.34	IRVVGVAPGVVNTDM
Ribitol DH Klebsiella	181	29.77	VRVGAVLPGPVVTAL
3-Oxoacyl-[Acyl-Carrier Prot]	174	41.42	ITVNVVAPGFIETDM

Figure 3.2   (a) continued.

MOTIF 7

```
                              bits 6.1
                                   5.5
                                   4.9
                                   4.2
                   Information     3.6
                   content         3.0
                   (24.6 bits)     2.4 *  *  *           *
                                   1.8 *  ***      *    *
                                   1.2 * *********
                                   0.6 ***********
                                   0.0 -----------

                   Multilevel          GRIINIASVAG
                   consensus                V  G
                   sequence                    S
```

Sequence name	Start	Score	Site
11beta-HSD-TYPE1	163	22.43	GSIVVVSSLAG
11beta-HSD-TYPE2	211	23.84	GRIVTVGSPAG
15-OH Prostaglandin DH	131	29.82	GIIINMSSLAG
20beta-HSD Streptomyces	132	29.45	GSIVNISSAAG
3beta-HSD Bacterial	130	26.51	GSIINMASVSS
ActIII protein Streptomyces	137	27.22	GRIVNIASTGG
7alpha-HSD Eubacterium	137	28.99	GVIINTASVTG
ADH Drosophila	133	28.98	GIICNIGSVTG
Ribitol DH Klebsiella	140	20.52	GDIIFTAVIAG
3-Oxoacyl-[Acyl-Carrier Prot]	131	27.42	GRIITIGSVVG

MOTIF 8

```
                              bits 6.1
                                   5.5
                                   4.9
                                   4.2
                   Information     3.6
                   content         3.0 *           *      *
                   (31.5 bits)     2.4 *           *      *
                                   1.8 * *         *    * *
                                   1.2 * *    **** ***
                                   0.6 * * **********
                                   0.0 ---------------

                   Multilevel          ExFxKVLEINLTGVF
                   consensus             W   I
```

Sequence name	Start	Score	Site
11beta-HSD-TYPE2	182	23.36	ATFRSCMEVNFFGAL
15-OH Prostaglandin DH	98	27.93	KNWEKTLQINLVSVI
20beta-HSD Streptomyces	102	40.51	ERFRKVVDINLTGVF
3beta-HSD Bacterial	101	28.96	EDFSRLLKINTESVF
ActIII protein Streptomyces	105	36.85	ELWLDVVETNLTGVF
ActIII protein Streptomyces	200	9.92	EHYSDIWEVSTEEAF
7alpha-HSD Eubacterium	107	38.29	EEFKHIMDINVTGVF
ADH Drosophila	221	27.90	ENFVKAIELNQNGAI
Ribitol DH Klebsiella	110	27.06	DVWDRVLHLNINAAF
3-Oxoacyl-[Acyl-Carrier Prot]	101	35.22	EEWNDIIETNLSSVF

Figure 3.2 (a) continued.

MOTIF 9

```
                                    bits 6.1
                                         5.5
                                         4.9
                                         4.2
                         Information     3.6
                         content         3.0   *         *
                         (32.9 bits)     2.4 * *        *    *
                                         1.8 * ***    *  **  **
                                         1.2 * ***    *  **  **
                                         0.6 * ************
                                         0.0 --------------

                         Multilevel          PEEIAEAVLFLASD
                         consensus            V N    A

Sequence name                       Start Score          Site
20beta-HSD Streptomyces              214  32.56      PGEIAGAVVKLLSD
3beta-HSD Bacterial                  219  35.23      PERIAQLVLFLASD
ActIII protein Streptomyces          229  26.34      PSEVAEMVAYLIGP
7alpha-HSD Eubacterium               218  39.59      PEEIANVYLFLASD
Ribitol DH Klebsiella                216  29.88      PIEVAESVLFMVTR
3-Oxoacyl-[Acyl-Carrier Prot]        212  39.45      AQEIANAVAFLASD
```

MOTIF 10

```
                                    bits 6.1
                                         5.5
                                         4.9   *       *
                                         4.2   *       *
                         Information     3.6   *       *
                         content         3.0   *       *
                         (35.4 bits)     2.4   *     ***
                                         1.8 * *   **** *
                                         1.2 * *  *******
                                         0.6 **********
                                         0.0 -----------
                         Multilevel          TPCLITGWGLT
                         consensus           S     S

Sequence name                       Start Score          Site
TRYPSIN BOVINE                       120  36.11       TQCLISGWGNT
CHYMOTRYPSIN BOVINE                  134  35.24       TTCVTTGWGLT
ELASTASE PIG                         151  38.61       SPCYITGWGLT
```

Figure 3.2   (a) continued.

MOTIF 11

```
                            bits 6.1
                                 5.5
                                 4.9                    *
                                 4.2                  * *
             Information        3.6                  * *
             content           3.0                  * *
             (31.6 bits)       2.4                  * *
                                 1.8        *  * ****
                                 1.2    *   *******
                                 0.6    **********
                                 0.0    -----------

             Multilevel             GSKIKDNMFCA
             consensus              TR T
```

Sequence name	Start	Score	Site
TRYPSIN BOVINE	159	29.99	PGQITSNMFCA
CHYMOTRYPSIN BOVINE	173	34.23	GTKIKDAMICA
ELASTASE PIG	191	28.91	GSTVKNSMVCA
THROMBIN BOVINE	544	34.38	RIRITDNMFCA

MOTIF 12

```
                            bits 6.1
                                 5.5
                                 4.9                       *
                                 4.2                       *
             Information        3.6                       *
             content           3.0    *                   *
             (43.2 bits)       2.4    **    *             *
                                 1.8    *** ****    *    **
                                 1.2    *******   **  **
                                 0.6    ***************
                                 0.0    ---------------

             Multilevel             KPGVYTRVSALVSWI
             consensus                 F   T  YI
```

Sequence name	Start	Score	Site
TRYPSIN BOVINE	208	49.02	KPGVYTKVCNYVSWI
CHYMOTRYPSIN BOVINE	224	40.73	TPGVYARVTALVNWV
ELASTASE PIG	245	45.90	KPTVFTRVSAYISWI
THROMBIN BOVINE	602	38.04	KYGFYTHVFRLKKWI

Figure 3.2 (a) continued.

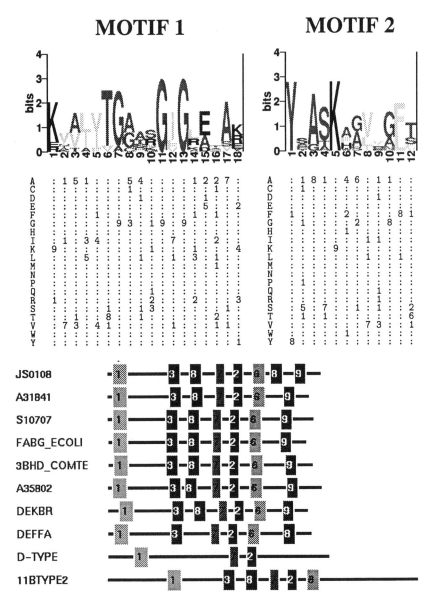

Figure 3.2 (b) Detailed views of two motifs discovered by MEME in the alcohol dehydrogenase training set and diagrams showing the positions of all motifs in 10 training set sequences. The diagrams at the top indicate the information content of the most informative letters in each column of the motif. (Letter height is equal to information content.) The matrix below each information diagram shows a low-resolution version of the letter-frequency matrix defining the motif. The frequency of each letter at each motif position is shown scaled by 10 and rounded to one digit, with ":" replacing zero for readability. The positions and spacings of motif occurrences in 10 sequences are indicated in the diagrams below the frequency matrices.

MAST ANALYSIS OF DEHYDROGENASE/PROTEASE TRAINING SET

SEQUENCE NAME	DESCRIPTION	E-VALUE	LENGTH
ACT3_STRCO	ActIII protein - Streptomyces coeli...	4.9e-64	261
BA72_EUBSP	27K-2 Protein (Cholic Acid-Induced)..	6.6e-63	249
2BHD_STREX	20beta-Hydroxysteroid Dehydrogenase. .	2.3e-62	255
FABG_ECOL	3oxoacyl-[Acyl-Carrier Protein] Re. .	1.8e-61	244
3BHD_COMTE	3beta-hydroxysteroid Dehydrogenase. ..	5.4e-54	253
PGDH_HUMAN	15Hydroxyprostaglandin Dehydrogena..	5.6e-53	266
RIDH_KLEAE	Ribitol 2-Dehydrogenase (EC 1.1.1.5...	1.1e-46	247
ADH_DROME	Alcohol Dehydrogenase (EC 1.1.1.1) ...	6.0e-31	256
DHI1_HUMAN	11beta HSD-Type 1-NADP Dependent-Hum.	2.7e-23	292
DHI2_HUMAN	11beta HSD-Type 2-NAD Dependent-Hum..	4.4e-21	404
CTRA_BOVIN	Chymotrypsinogen A (EC 3.4.21.1)	0.0017	245
EL1_PIG	Elastase 1 Precursor (EC 3.4.21.36)	0.91	266
TRYP_BOVIN	Trypsinogen (EC 3.4.21.4)	2.5	229
THRB_BOVIN	Prothrombin Precursor (EC 3.4.21.5)	7.2	625

MOTIF DIAGRAMS

SEQUENCE NAME	E-VALUE	MOTIF DIAGRAM
ACT3_STRCO	4.9e-64	6-[1]-54-[3]-11-[8]-17-[7]-9-[2]-11-[6]-34-[9]-19
BA72_EUBSP	6.6e-63	6-[1]-56-[3]-11-[8]-15-[7]-9-[2]-11-[6]-23-[9]-18
2BHD_STREX	2.3e-62	6-[1]-51-[3]-11-[8]-15-[7]-9-[2]-11-[6]-24-[9]-28
FABG_ECOLI	1.8e-61	5-[1]-51-[3]-11-[8]-15-[7]-9-[2]-11-[6]-23-[9]-19
3BHD_COMTE	5.4e-54	6-[1]-50-[3]-11-[8]-14-[7]-9-[2]-14-[6]-28-[9]-21
PGDH_HUMAN	5.6e-53	5-[1]-56-[3]- 3-[8]-18-[7]-9-[2]-13-[6]-31-[9]-31
RIDH_KLEAE	1.1e-46	14-[1]-51-[3]-11-[8]-15-[7]-7-[2]-11-[6]-20-[9]-18
ADH_DROME	6.0e-31	7-[1]-56-[3]-36- [7]-9-[2]-11-[6]-66
DHI1_HUMAN	2.7e-23	34-[1]-110- [7]-9-[2]-98
DHI2_HUMAN	4.4e-21	81-[1]-55-[3]-12-[8]-14-[7]-9-[2]-11-[6]-136
CTRA_BOVIN	0.0017	176-[1]-51
EL1_PIG	0.91	19-[2]-235
TRYP_BOVIN	2.5	229
THRB_BOVIN	7.2	489-[2]-124

Figure 3.3 MAST analysis of the training set with motifs for dehydrogenases. *E* value is the expected number of sequences in the database with an equal or higher score.

cause of their well-developed mathematical theory and the existence of computationally efficient HMM training and searching algorithms. However, standard protein HMMs have at least two major weaknesses. First, they employ a linear topology that implicitly assumes an oversimplified model of molecular evolution that allows only point mutations, insertions, and deletions. Second, standard protein HMMs are too large. A typical such model contains on the order of 5000 parameters. To estimate the values of these parameters reliably, a training set of dozens of proteins that are already known to be related is required. Therefore, current HMMs are inadequate for modeling newly discovered or conjectured protein families for which only a few members are known.

Meta-MEME is an extension of MEME into an HMM framework. A set of gapless MEME motifs characterizing a single family of proteins is combined in a single linear HMM. A Meta-MEME model typically contains approximately one-sixth as many parameters as a corresponding standard HMM. Therefore, Meta-MEME models can be trained more quickly and with smaller training sets. Furthermore, Meta-MEME's focus on highly conserved regions allows for the detection of more remote homologies.

3.3   Results

As discussed above, to examine the selectivity and sensitivity of MEME, Meta-MEME, and MAST with small datasets, we use a training set of 10 diverse dehydrogenases and four serine proteases, which belong to different protein superfamilies. We examine how effectively MEME analyzes dehydrogenases in the presence of four proteins unrelated to the dehydrogenases. Conversely, we determine if MEME can find important motifs in the four serine proteases in the presence of 10 dehydrogenases. The SWISS-PROT identifiers and the title lines for the sequences in the training dataset are shown in table 3.1.

Figure 3.2(a) shows the first 12 motifs identified by MEME in the dataset. The so-called "multilevel" consensus summary shows which individual residues appear with at least 20% probability in each position of the instances of a motif, with the highest probability amino acid above and lower probability amino acids below. The information content (measured in bits) on the ordinate reveals the degree of conservation of the amino acid(s) at each position in the consensus sequence. For each subsequence of any protein that resembles the consensus motif, the protein name and its subsequence are printed, with a numerical measure of how well the subsequence resembles the consensus motif.

As shown in figure 3.2(a), only the dehydrogenases have the first three motifs while only the serine proteases have motifs 4 and 5; only dehydrogenases have motifs 6–9, and only serine proteases have motifs 10–12. This separation shows that MEME can take 14 sequences from two protein families and sort out each family.

Table 3.1 SWISS-PROT identifiers and descriptions for the training set.

Identifier	Description
2BHD_STREX	20-β-Hydroxysteroid dehydrogenase
3BHD_COMTE	3-β-Hydroxysteroid dehydrogenase
ACT3_STRCO	Putative ketoacyl reductase
ADH_DROME	Alcohol dehydrogenase
BA72_EUBSP	7-α-Hydroxysteroid dehydrogenase
DHI1_HUMAN	Corticosteroid 11-β-dehydrogenase type 1
DHI2_HUMAN	Corticosteroid 11-β-dehydrogenase type 2
FABG_ECOLI	3-Oxoacyl-[acyl-carrier protein] reductase
PGDH_HUMAN	15-Hydroxyprostaglandin dehydrogenase [NAD(+)]
RIDH_KLEAE	Ribitol 2-dehydrogenase
CTRA_BOVIN	Chymotrypsinogen A (EC 3.4.21.1)
EL1_PIG	Elastase 1 precursor (EC 3.4.21.36)
TRYP_BOVIN	Trypsinogen (EC 3.4.21.4)
THRB_BOVIN	Prothrombin precursor (EC 3.4.21.5)

3.3.1 Motif Analysis

Here we discuss the first five motifs in detail to show how MEME extracts important information that can elucidate functional properties of the proteins in the dataset.

Motif 1 is 18 amino acids long and corresponds to the β-strand, turn, α-helix that is common to dehydrogenases (Wierenga et al., 1985; Branden and Tooze, 1991). This structure is part of the binding site for the nucleotide cofactor NAD(P)(H). Note the high scores, which begin at 27 bits and go to 51 bits, indicating that this is a very strong motif for the dataset.

Motif 2 is 12 amino acids long and contains the tyrosine and lysine residues that are important in catalysis. Both tyrosine and lysine have high information content scores, as does a phenylalanine residue. Here, too, the dehydrogenases have high scores, with the interesting exception of 11β-HSD type 2 enzyme, which has a score of 17 bits. This score is significantly above the cutoff, but it is below that of the others. This indicates special properties of this part of 11β-HSD type 2 and shows how MEME can provide important clues, in a quantitative fashion, about specific enzymes in a family.

Motif 3 is 15 amino acids long and is a highly conserved segment that binds the nucleotide cofactor. The 11β-HSD type 2 enzyme has a score of only 9 bits, and the type 1 enzyme is below the cutoff of 8. This indicates that this motif is unusual in both of these steroid dehydrogenases.

Motif 4 is 17 amino acids long and is found in only the four serine proteases. This motif contains the histidine residue at the catalytic site. This histidine along with a serine and aspartic acid form the catalytic triad that is characteristic of serine proteases.

Motif 5 is 11 amino acids long and is found in only the four serine proteases. This motif contains the serine residue at the catalytic site.

Description of the first five motifs shows that MEME examines the 14 sequences in the training set and identifies a motif characteristic of the 10 dehydrogenases as most important in the dataset. The information in the 10 dehydrogenases dominates the MEME analysis for the next two motifs. However, after three motifs, the information in the serine proteases is strong enough that these proteases are the sequences that MEME parses for motifs 4 and 5. Then MEME finds that the next four strongest motifs are in the 10 dehydrogenase sequences and the last three motifs are from serine proteases. This demonstrates how MEME outputs motifs in a hierarchical order and can extract meaningful information from four sequences even if they are embedded in a larger set of sequences to which they are unrelated.

3.3.2   Using MEME Motifs in the MAST Searching Tool

One application of the motifs generated by MEME from a set of related proteins is to search databases and identify distant unrecognized homologs that would not be found in a routine BLAST or Fasta search.

Another important application is to examine a newly sequenced open reading frame (ORF) for high-scoring MEME-generated motifs, as is done now with PROSITE signatures, to identify the ORF's ancestry and functional domains. In comparison to PROSITE signatures, MEME offers the flexibility to search an ORF against several alternative motifs for a single protein family. In the short-chain alcohol dehydrogenase family, the PROSITE signature uses the first five residues of motif 2 because these positions contain the highly conserved tyrosine and lysine. As we showed previously (Bailey et al., 1997), this motif is not the most characteristic motif for short-chain alcohol dehydrogenases. A longer motif better describes this part of the dehydrogenases. Moreover, some members of the short-chain alcohol dehydrogenase family do not conserve either the tyrosine or lysine (Baker, 1995). Others, such as protocholorophyllide reductase (Baker, 1994b) and *Myxococcus xanthus* C-factor (Baker, 1994a), have residues adjacent to the tyrosine that disagree with the PROSITE signature. The MAST tool uses all of the motifs in its search.

MAST Analysis of the Training Set

As an additional demonstration of the selectivity of MEME, we used MAST to search the training set with the seven dehydrogenase and five serine protease motifs. Figures 3.3 and 3.4 show the MAST analysis of the training set with a probe of the dehydrogenase motifs (1–3, 6–9) and a probe of the serine protease motifs (4, 5, and 10–12). The scores clearly show that the first set of motifs are characteristic of the dehydrogenases and the second set are characteristic of the serine proteases, demonstrating that the MEME analysis correctly identifies motifs specific for proteases and dehydrogenases. The motifs specific to each family do not appear with statistical significance in proteins belonging to the other family.

An important observation is that the order and spacing of the motifs are conserved in each family. This property is an important basis for the high selectivity of the Meta-MEME tool, which uses HMMs to construct a probe for searching sequence databases, as described in section 3.2.2 of this chapter.

SEQUENCE NAME	DESCRIPTION	E-VALUE	LENGTH
TRYP_BOVIN	Trypsinogen (EC 3.4.21.4)	1.5e-54	229
CTRA_BOVIN	Chymotrypsinogen A (EC 3.4.21.1)	5.6e-54	245
EL1_PIG	Elastase 1 Precursor (EC 3.4.21.36)	1.5e-53	266
THRB_BOVIN	Prothrombin Precursor (EC 3.4.21.5)	4.4e-37	625
ACT3_STRCO	ActIII protein - Streptomyces coeli...	1.5	261
3BHD_COMTE	3Beta-Hydroxysteroid Dehydrogenase. ..	1.9	253
DHI1_HUMAN	11beta HSD-Type 1-NADP Dependent-Hum..	2.6	292
DHI2_HUMAN	11BETA HSD-Type 2-NAD Dependent-Hum...	4.5	404
2BHD_STREX	20beta-Hydroxysteroid Dehydrogenase...	5.5	255
BA72_EUBSP	27K-2 Protein (Cholic Acid-Induced)...	5.7	249
ADH_DROME	Alcohol Dehydrogenase (EC 1.1.1.1) ...	7.9	256
FABG_ECOLI	3-Oxoacyl-[Acyl-Carrier Protein] Re...	8.6	244
PGDH_HUMAN	15-Hydroxyprostaglandin Dehydrogena...	8.9	266

SEQUENCE NAME	E-VALUE	MOTIF DIAGRAM
TRYP_BOVIN	1.5e-54	30-[4]-72-[10]-28-[11]-9- [5]-18-[12]-7
CTRA_BOVIN	5.6e-54	41-[4]-75-[10]-28-[11]-7- [5]-22-[12]-7
EL1_PIG	1.5e-53	55-[4]-78-[10]-29-[11]-8- [5]-24-[12]-7
THRB_BOVIN	4.4e-37	393-[4]-89-[10]-33-[11]-12-[5]-24-[12]-9
ACT3_STRCO	1.5	176-[11]-74
3BHD_COMTE	1.9	127-[4]-109
DHI1_HUMAN	2.6	292
DHI2_HUMAN	4.5	404
2BHD_STREX	5.5	255
BA72_EUBSP	5.7	249
ADH_DROME	7.9	256
FABG_ECOLI	8.6	244
PGDH_HUMAN	8.9	266

Figure 3.4   MAST analysis of the training set with motifs for serine proteases.

Identifying Homologs in Genpept98 with MAST

We searched Genpept98 with the seven motifs characteristic of the dehydrogenases and the five characteristic of the serine proteases. A histogram of the output is shown in figures 3.5 and 3.6. Both probes identify several hundred homologs. The minima in the output indicate a good separation of true positives from other sequences. In figure 3.5, although only 10 dehydrogenases are in the training set, MAST identifies over 600 homologs with excellent separation from the over 200,000 proteins in Genpept98. In figure 3.6, although only four serine proteases are in the training set, MAST identifies over 600 homologs with excellent separation from the over 200,000 proteins in Genpept98. The output of predicted homologs is bimodal. The second peak around a score of 10 is due to haptoglobins, which do not have proteolytic activity though they are homologs of serine proteases.

The MAST output identifies homologs that are distantly related to those in the dataset. These distantly related proteins can be added to the training set and analyzed by MEME again. Then MAST can search Genpept with the motifs from this analysis. In this way, one can easily build up a training set that is characteristic of the protein family.

3.3.3   Results from Meta-MEME

As mentioned above, an important property of MEME motifs is that their order and spacing are conserved, as shown in figures 3.3 and 3.4. Conservation of motif order and spacing is biologically reasonable because the motifs correspond to functional structures. For example, consider the spatial relationship in the dehydrogenases between motif 1, where the glycine-rich turn is at the amino-terminus, and motif 2, which contains the catalytically important residues. A protein with these motifs in the reverse order would not be homologous in its 3D structure to the short-chain alcohol dehydrogenases. Thus, the order of the motifs is a meta-motif or fingerprint for a given protein superfamily. In this way, even a motif with a low score that is present in the canonical order provides information that improves confidence in assigning the unknown protein to a superfamily.

Meta-MEME HMM probes for dehydrogenases and proteases were constructed and used to search Genpept98. As shown in figures 3.7 and 3.8, Meta-MEME identifies distantly related homologs of each family, with excellent separation from unrelated proteins. In figure 3.7, using a probe consisting of the seven motifs for dehydrogenases in their proper order and spacing, Meta-MEME identifies over 600 dehydrogenase homologs with excellent separation from the nonhomologous proteins in Genpept98. In figure 3.8, although only four serine proteases are in the training set, Meta-MEME identifies over 600 homologs with excellent separation from the over 200,000 proteins in Genpept98. The output of predicted homologs is bimodal. The second peak with E values around 10^{-6} is again due to haptoglobins.

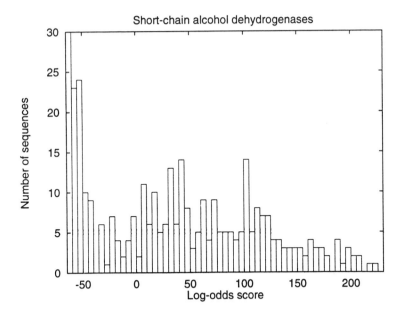

Figure 3.5 Histogram of MAST search of Genpept98 with motifs for dehydrogenases.

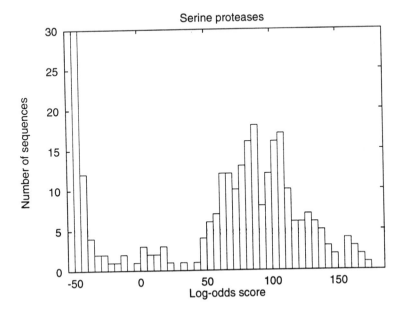

Figure 3.6 Histogram of MAST search of Genpept98 with motifs for serine proteases.

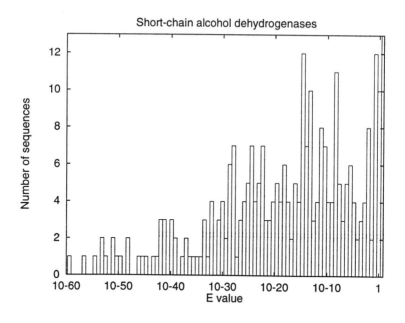

Figure 3.7 Histogram of Meta-MEME search of Genpept98 with motifs for dehydrogenases.

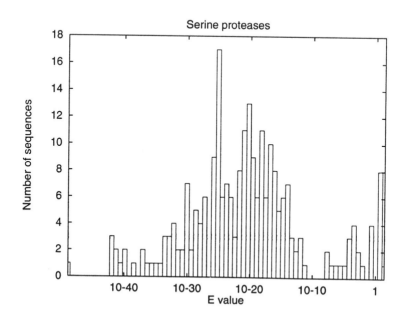

Figure 3.8 Histogram of Meta-MEME search of Genpept98 with motifs for serine proteases.

3.3.4   MEME Motifs for Local Sequence Alignment

With the increased size of protein families as many genomes are sequenced in their entirety, there is increased interest in constructing multiple alignments for functional and phylogenetic analyses. We have found that MEME motifs can give a good phylogenetic analysis of dehydrogenases (Grundy et al., 1997).

Recently, we have also found that MEME motifs are useful for local sequence alignments of diverse datasets. Indeed, experiments show that without any additional modification, MEME yields a better global multiple alignment based on conserved regions than all but the Feng-Doolittle DFALIGN method (Feng and Doolittle, 1987), which MEME approaches in accuracy. In the experiments no nonhomologous proteins were included in the training sets, as to do so would bias the comparison against progressive alignment methods, which always align all members of a training set, as do other alignment methods, unlike MEME.

The similar accuracy of MEME and DFALIGN is intriguing because the methods approach the alignment problem from opposite directions. MEME treats all sequences similarly and simultaneously, while DFALIGN aligns the closest sequences and progressively adds the next closest sequence to the alignment. We are now working to incorporate additional strategies into a new alignment tool called MEME-Align, which we think can be widely used for complex datasets of distantly related proteins.

3.4   Discussion

We have used an unsupervised learning algorithm, MEME, to analyze a diverse group of proteins from two different protein families. MEME successfully identifies motifs specific to each protein family. The output for the 10 dehydrogenases is not compromised by inclusion of four serine proteins. Moreover, the serine protease motifs are also accurate and informative. Both motifs are sensitive probes for the MAST and Meta-MEME database-searching tools, even without bootstrapping the output into a dataset for further analysis by MEME.

Our earlier studies with MEME and MAST demonstrated their impressive sensitivity for identifying distantly related homologs (Bailey et al., 1997). For example, MAST identified yeast UDP-glucose-4-epimerase as a member of the short-chain alcohol dehydrogenase family in agreement with an analysis of the 3D structure of UDP-glucose-4-epimerase and 20β-HSD and dihydropteridine reductase. Database searches with BLAST or Fasta do not find this homology. MEME and MAST are the first sequence analysis programs to give statistical evidence that the UDP-epimerase protein family (which includes mammalian 3β-HSD plant dihydroflavonol reductase, and bacterial cholesterol dehydrogenase) is homologous to short-chain dehydrogenases.

MAST identifies several proteins (*Mycobacterium tuberculosis* [InhA], *Escherichia coli* enoyl-acyl-carrier protein reductase [EnvM], rat 2,4-dienoyl-coacyl reductase, and *Saccharomyces cerevisiae* sporulation-specific protein [SPX19]) as homologs of HSDs. These distantly related homologs do not contain the signature motif for the short-chain alcohol dehydrogenase family, which prevents identification of their ancestry with PROSITE.

The identification of these distant homologs highlights an important advantage of extracting several motifs for a protein superfamily: even if a sequence lacks one motif, its ancestry can still be identified from an analysis of the other motifs it possesses. In other words, MEME can identify homologies in protein sequences that do not all contain the same motifs. In the case of *Mycobacterium tuberculosis* InhA, a target for drugs such as isoniazid and ethionamide for controlling tuberculosis (Baker, 1995), this can have important medical applications.

The Meta-MEME tool is even more powerful than MAST. A search of Genpept with Meta-MEME (Grundy et al., 1997) identifies many more sugar epimerases as belonging to the short-chain alcohol dehydrogenase family, as well as other proteins whose ancestry cannot be identified by conventional BLAST searches. For example, *Cornyebacterium halohydrin* epoxidases A and B, which were entered into GenBank without noting their homology to dehydrogenases, are identified as homologs of HSDs. Moreover, Meta-MEME clearly shows that dihydropteridine reductase is homologous to HSDs, despite the low sequence similarity of dihydropteridine reductase to short-chain alcohol dehydrogenases.

By examining the output from a database search for the order and spacing of all the motifs jointly, one can identify distantly related homologs among the noise of proteins that by chance have a positive score for a single motif. The motifs found by MEME correspond to matching 3D features in those dehydrogenases whose 3D structure is known, so the organization of the motifs is a mapping of 3D information onto the one-dimensional sequence information. This may explain the sensitivity of the meta-motif analysis.

An important global feature of the MEME meta-motif output is that it provides information on the similarities and differences in individual positions and motifs for a protein family. In this way, MEME elucidates conserved domains that are likely to have similar functions in protein family as well as idiosyncratic domains that are likely to be important in the unique properties of an enzyme, as we showed for 11β-HSD and 17β-HSD (Bailey et al., 1997). This information can be used for mutagenesis studies and in an analysis of their tertiary structure to begin to elucidate the mechanism of action of these proteins.

Chapter 4

Pattern Discovery and Classification in Biosequences

Jason T. L. Wang, Thomas G. Marr, Steve Rozen,
Dennis Shasha, Bruce A. Shapiro, Gung-Wei Chirn,
Zhiyuan Wang, and Kaizhong Zhang

With the significant growth of the amount of biosequence data, it becomes increasingly important to develop new techniques for finding "knowledge" from the data. *Pattern discovery* is a fundamental operation in such applications. It attempts to find patterns in biosequences that can help scientists to analyze the property of a sequence or predict the function of a new entity. The discovered patterns may also help to *classify* an unknown sequence, that is, assign the sequence to an existing family. In this chapter, we show how to discover active patterns in a set of protein sequences and classify an unlabeled DNA sequence. We use protein sequences as an example to illustrate our discovery algorithm, though the algorithm applies to sequences of any sort, including both protein and DNA.

4.1 Pattern Discovery in Protein Sequences

4.1.1 Preliminaries

The patterns we wish to discover within a set of sequences are regular expressions of the form $*X_1 * X_2 * \ldots$. The $X_1, X_2, \ldots$ are *segments* of a sequence, that is, subsequences made up of consecutive letters, and $*$ represents a variable length don't care (VLDC). In matching the expression $*X_1 * X_2 * \ldots$ with a sequence S, the VLDCs may substitute for zero or more letters in S at zero cost.

The dissimilarity measure used in comparing two sequences is the *edit distance*, that is, the minimum cost of edit operations used to transform one

S_1: YDPMIEDKEYSRLVG

S_2: RMKQLGRTYDPAVWG

S_3: YDPMNWNFEKETLVG

Figure 4.1   The set S of three sequences.

subsequence to the other after an optimal and zero-cost substitution for the VLDCs, where the edit operations include insertion, deletion, and change of one letter to another (Wagner and Fischer, 1974; K. Zhang et al., 1994). That is, we find a one-to-one mapping from each VLDC to a subsequence of the data sequence and from each pattern subsequence to a subsequence of the data sequence such that the following two conditions are satisfied. (i) The mapping preserves the left-to-right ordering (if a VLDC at position i in the pattern maps to a subsequence starting at position i_1 and ending at position i_2 in the data sequence, and a VLDC at position j in the pattern maps to a subsequence starting at position j_1 and ending at position j_2 in the data sequence, and $i < j$, then $i_2 < j_2$). (ii) The mapping has the minimum cost among all such mappings, where the cost of a mapping is the sum of the costs from pattern subsequences to data subsequences in the mapping. The edit distance is a useful measure of evolutionary distance (Sankoff and Kruskal, 1983). For the purpose of this work, we assume that all the edit operations have a unit cost, though the techniques we propose do not depend on that cost assumption or essentially on the edit distance metric.

Example 4.1 Consider the set S of three sequences in figure 4.1. Suppose only exactly coinciding segments occurring in at least two sequences and having lengths greater than 3 are considered "active." Then S contains one active pattern:

$$*S_1[1,4]* = *\text{YDPM}* \iff *S_3[1,4]* = *\text{YDPM}*,$$

where $V[x,y]$ is a segment of a sequence V from the xth to the yth letter inclusively. If patterns occurring in all the three sequences within distance one are considered active, that is, if one mutation (mismatch, insertion, or deletion) is allowed in matching a pattern with a sequence, then S contains three active patterns:

$$*S_1[1,4]* = *\text{YDPM}*$$
$$\iff *S_2[8,11]* = *\text{TYDP}*$$
$$\iff *S_2[9,12]* = *\text{YDPA}*$$
$$\iff *S_3[1,4]* = *\text{YDPM}*$$

If patterns having the form $*X*Y*$ are sought with lengths greater than 7 and one mutation allowed, then S_1 and S_3 share the following four active patterns:

$$*S_1[1,4] * S_1[12,15]* = *\texttt{YDPM} * \texttt{RLVG} *$$
$$\Longleftrightarrow \quad *S_1[1,5] * S_1[13,15]* = *\texttt{YDPMI} * \texttt{LVG} *$$
$$\Longleftrightarrow \quad *S_3[1,4] * S_3[12,15]* = *\texttt{YDPM} * \texttt{TLVG} *$$
$$\Longleftrightarrow \quad *S_3[1,5] * S_3[13,15]* = *\texttt{YDPMN} * \texttt{LVG}*$$

Formally, let S be a set of sequences. We define the occurrence number (or the activity) of a pattern as the number of sequences in S that match the pattern within the allowed distance. We say the occurrence number of a pattern P with respect to distance i and set S, denoted $occurrence\_no_S^i(P)$, is k if $*P*$ matches k sequences in S within distance i; that is, the k sequences contain P within distance i. For example, in figure 4.1, $occurrence\_no_S^0(*\texttt{YDPM}*) = 2$ and $occurrence\_no_S^1(*\texttt{YDPM}*) = occurrence\_no_S^1(*\texttt{TYDP}*) = occurrence\_no_S^1(*\texttt{YDPA}*) = 3$.

Given a set of sequences S, we want to find all the active patterns P where P is within the allowed distance $Dist$ of at least $Occur$ sequences in S and $|P| \geq Length$, where $|P|$ represents the number of the non-VLDC letters in the pattern P. ($Dist$, $Occur$, $Length$, and the form of P are user-specified parameters.) The basic subroutine is to match a given pattern with a sequence in the given set. For example, in matching $*\texttt{TQI}*$ with a sequence $\texttt{MYALTIHKR}$, the first asterisk would substitute for $\texttt{MYAL}$ and the second asterisk would substitute for $\texttt{HKR}$. The distance is 1 (representing the cost of deleting $\texttt{Q}$). The length of the pattern $*\texttt{TQI}*$ is 3. Notice that given a regular expression pattern P and sequence S, one can determine whether P is within distance $Dist$ of S in $O(Dist \times |S|)$ time when $O(|P|) = O(\log |S|)$ (Wu and Manber, 1992).

To discover active patterns in a set of sequences, our overall strategy is first to find candidate segments among a small sample and then to combine the segments into candidate patterns. We then check which patterns satisfy the specified requirements. Many techniques have been published previously to solve similar problems. These problems are mostly concerned with discovering patterns made up of single segments, or multiple segments separated by fixed length don't cares. One commonly used technique is based on multiple sequence alignment (see Waterman, 1989, for review). The technique is useful when entire sequences in the set are similar. However, when the sequences have only short regions of local similarities, this approach makes no sense. There are also techniques based on local similarity search. The techniques work effectively when similarities meet some constraints, such as occurring in a predetermined number of sequences in the set (Roytberg, 1992), differing by mismatches but not by insertions/deletions (Bacon and Anderson, 1986), or being situated at almost the same distance from the start of the sequences (Vingron and Argos, 1989). In contrast to these techniques, our approach can find similarities composed of nonconsecutive segments separated by VLDCs without prior knowledge of their structures, positions, or occurrence frequency.

4.1.2   Discovery Algorithm

Our algorithm consists of two phases: (1) find candidate segments among a small sample $\mathcal{A}$ of the sequences, and (2) combine the segments to form candidate patterns and evaluate the activity of the patterns in all of $\mathcal{S}$ to determine which patterns satisfy the specified requirements.

Phase 1 consists of two subphases. In subphase A, we construct an index structure for the sequences in the sample. In subphase B, we traverse the structure to locate the candidate segments.

Subphase A of Phase 1

We construct a *generalized suffix tree* (GST; Hui, 1992) for the sample of sequences. A suffix tree is a trie-like data structure that compactly represents a string by collapsing a series of nodes having one child to a single node whose parent edge is associated with a string (McCreight, 1976; Landau and Vishkin, 1989). A GST is an extension of the suffix tree, designed for representing a set of strings. Each suffix of a string is represented by a leaf in the GST. Each leaf is associated with an index i. The edges are labeled with character strings such that the concatenation of the edge labels on the path from the root to the leaf with index i is a suffix of the ith string in the set. Figure 4.2 gives an example (the node labeled 1 above the leaf MTRM is an example of the result of a collapsing).

Informally, the algorithm for constructing the GST works as follows. We append a unique symbol to each sequence in the sample and concatenate the sequences in the sample into a single one. We insert the suffixes of the sequences as into a trie except that if a node has only one child, we collapse the child with the parent and label the edge going down from the parent with a substring instead of a single character. The GST can be constructed asymptotically in $O(n)$ time and space where n is the total length of all sequences in the sample $\mathcal{A}$.

Subphase B of Phase 1

In subphase B, we traverse the GST constructed in subphase A to find all segments (i.e., all prefixes of strings labeled on root-to-leaf paths) that satisfy the length minimum. If the pattern specified by the user has the form $*X*$, then the length minimum is simply the specified minimum length of the pattern. If the pattern specified by the user has the form $*X_1 * X_2*$, we find all the segments V_1, V_2 where at least one of the V_i, $1 \leq i \leq 2$, is larger than or equal to half of the specified length and the sum of their lengths satisfies the length requirement. If the user-specified pattern has the form $*X_1 * X_2 * \ldots * X_k*$, we find the segments $V_1, V_2, \ldots, V_k$ where at least one of the V_i, $1 \leq i \leq k$, is larger than or equal to $1/k$th of the

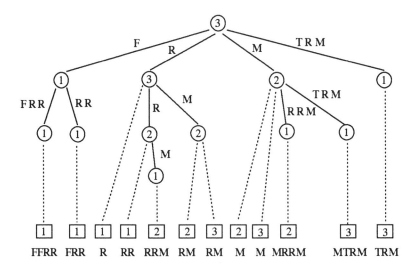

Figure 4.2 The GST for a sample $\mathcal{A} = \{$FFRR, MRRM, MTRM$\}$. Leaves are represented as rectangles, labeled with the indexes. Nonleaf nodes are represented as circles, labeled with the *count* values. The count value of a nonleaf node v represents the number of different indexes associated with the leaves in the subtree rooted at v. The suffix corresponding to a leaf is shown below the leaf. Note that the suffixes RM and M appear in two strings and hence appear twice in the leaves.

specified length and the sum of the lengths of all these segments satisfies the length requirement.

Phase 2

This phase also has two subphases. In subphase A, we evaluate the activity of the candidate patterns and rank them from highest to lowest according to their occurrence numbers in the sample with respect to distance $Dist$. If interesting patterns are of the form $*X_1 * X_2 * \ldots$, we consider all possible combinations $V_1, V_2, \ldots$ of the segments obtained in phase 1 that meet the length requirement and match $*V_1 * V_2 * \ldots$ with the sequences in the sample. Subphase B evaluates the most likely candidate patterns found in subphase A with respect to the entire set. (As our experimental results show [see section 4.1.4], screening out those unlikely candidate patterns in the first subphase saves significant time in the overall computation.)

4.1.3 Optimization Heuristics

In phase 2 of the discovery algorithm, we compare only the most likely candidate patterns with the entire set. The main question from an optimization point of view is which candidates to compare. Our strategy is as follows.

We use *simple random sampling without replacement* (Cochran, 1977) to select sample sequences from the set. Consider a candidate pattern P. Let M (a, respectively) denote the number of sequences in the entire set S (the sample $\mathcal{A}$, respectively) that contain P within the allowed distance. Let N be the set size and n the sample size; $F = M/N$ and $f = a/n$. Then, with probability = 99%, F is in the interval $(\hat{F}_L, \hat{F}_U)$ where

$$\hat{F}_L = f - (t\sqrt{\frac{N-n}{N-1}}\sqrt{\frac{f(1-f)}{n}} + \frac{1}{2n}),$$

$$\hat{F}_U = f + (t\sqrt{\frac{N-n}{N-1}}\sqrt{\frac{f(1-f)}{n}} + \frac{1}{2n}).$$

The symbol t is the value of the normal deviate corresponding to the desired confidence probability. When the probability = 99%, $t = 2.58$ (Cochran, 1977). The values of N, n are given; f, a can be obtained from subphase A of phase 2. Thus, if the estimator $(\hat{F}_U \times N) < Occur$ for the candidate pattern P, then with probability $\geq$ 99%, P will not be an active pattern satisfying the specified requirements. We therefore discard it. We refer to this pruning as *candidate pattern optimization*.

The second optimization heuristic we implemented is to eliminate the redundant calculation of occurrence numbers. Observe that the most expensive operation in our discovery algorithm is finding the occurrence number of a pattern with respect to the entire set, since that entails matching the pattern with all sequences. We say $*U_1 * \ldots * U_m*$ is a *subpattern* of $*V_1 * \ldots * V_m*$ if U_i is a subsegment of V_i, for $1 \leq i \leq m$.

One can observe that if P is a subpattern of P', then $occurrence\_no_S^k(P) \geq occurrence\_no_S^k(P')$ for any distance parameter k. Thus, if P' is in the final output set, we need not bother matching P with sequences in S, since P will be in the final output set, too. If P is not in the final output set, then P' will not be either, since its occurrence number will be even lower. We refer to this pruning strategy as *evaluation minimization*.

To illustrate how the above two optimization heuristics are incorporated into the discovery algorithm, consider finding the patterns of the form $*X * Y*$ whose total length is $\geq$ 5. We begin by enumerating segments of length 3 in the GST. Let $string(v)$ be the string on the edge labels from the root to v. If the above statistical estimator tells us that the combination of a segment $string(u_1)$ of length 3 with another segment $string(u_2)$ does not yield a pattern active enough to satisfy the specified $Dist$ and $Occur$ requirements, then we eliminate the pair $string(v_1)$ and $string(v_2)$ from consideration, where v_1 and v_2 are descendants of u_1 and u_2, respectively, in the GST. Similar pruning operations can be applied when enumerating longer segments in the GST.

Table 4.1 Experimental parameters and base values used in performance analysis.

Parameter	Value	Description
DSSize	150	Number of sequences in a dataset
NumSample	1	Number of samples tested for a dataset
SizeRatio	20%	Ratio between sample size and dataset size
Length	5	Minimum length of an interesting pattern
Dist	1	Allowed distance between a pattern and a sequence

4.1.4 Experiments and Results

We have carried out a series of experiments to evaluate the effectiveness and speed (measured by elapsed CPU time) of our approach. The data was a set of randomly generated 150 sequences, each having length 100. Every letter of the generated sequences was drawn randomly from the protein alphabet. For comparison purposes, we also tested the algorithms on real protein sequences. We randomly selected 150 proteins from the functionally related kinase family obtained from the Cold Spring Harbor Laboratory. The lengths of the kinase sequences ranged from 10 to 2938.

Table 4.1 shows the parameters and base values used in the experiments. The sequences in the sample were chosen randomly from the dataset. The parameter *NumSample* indicates the number of samples chosen for each dataset. In all the experiments presented here, only one sample was used in running a dataset. The sample size was obtained by multiplying *DSSize* by *SizeRatio*. The patterns of interest had the form $*X*Y*$.

The metric used to evaluate the effectiveness of our algorithms is

$$HitRatio = \frac{NumDiscovered}{TotalNum} \times 100\%,$$

where *NumDiscovered* is the number of interesting patterns discovered by our techniques. *HitRatio* stands for the percentage of the interesting patterns obtained from the exhaustive search method. The method works by considering all combinations of the segment pairs V_1, V_2 appearing in the dataset. One would like this percentage to be as high as possible. Notice that we have rejected approximately occurring patterns that never appear in the dataset, yet satisfy the *Dist* and *Occur* constraints, in favor of those that obey the constraints and do appear in the dataset. This is a theoretical limitation of our work that we have introduced to save computation time, though this also seems pragmatically to be a reasonable approach.

Figure 4.3 shows the effectiveness of our approach for varying sample sizes. In this experiment, we had turned on both candidate pattern optimization and evaluation minimization when running our algorithms. The minimum occurrence number required, *Occur*, was set to 60 for the artificial data and 8 for the kinase family. (The different parameter values were chosen to illustrate different results using different data.) Examin-

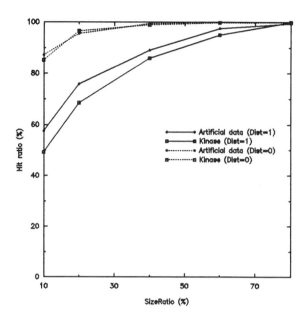

Figure 4.3   Effect of sample size.

ing the graphs, we see that when $Dist = 0$ and $SizeRatio \geq 0.2$, our approach behaves almost like an exhaustive search. When $Dist = 1$, the hit ratio reaches 80% provided the $SizeRatio \geq 0.4$. We were somewhat disappointed that smaller samples did not give a better hit ratio.

We next compared the running times of the algorithms for the $Dist = 1$ case. Figure 4.4 shows the results: our algorithms run significantly faster than the brute force method. Even with $SizeRatio = 0.8$, in which case the algorithms achieve nearly a 100% hit ratio, they are 10 times faster than an exhaustive search. When the sample is this large, both segments V_1, V_2 in an active pattern appear in the sample. Our algorithms work by enumerating all promising segment pairs in the sample and therefore can find all the interesting patterns.

We also examined the effectiveness of the proposed optimization heuristics. To isolate the effect of the heuristics, we started by turning off the optimizations, then turning on only one of them, and finally turning on both. To make the experiment manageable, we considered only patterns of the form $*X*$. The minimum occurrence number required, $Occur$, was set to 55. The other parameters had the values shown in table 4.1.

Figures 4.5 and 4.6 show the results obtained from the kinase sequences. (The results for the generated sequences are omitted since they lead to similar conclusions.) Examining the graphs, we see that very few active patterns were missed by the candidate pattern optimization. Pruning based on subpattern information works more effectively than that based on statistical

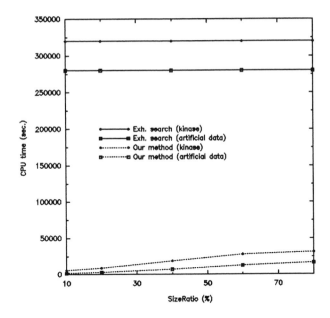

Figure 4.4 Comparison of running time.

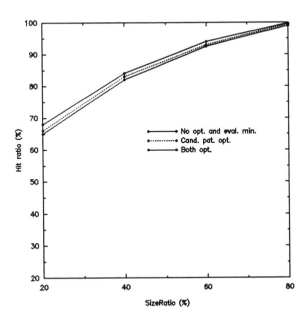

Figure 4.5 Performance of the pruning techniques.

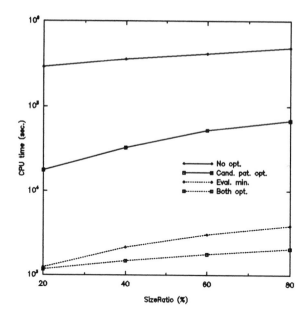

Figure 4.6   Efficiency of the pruning techniques.

estimation. Both optimizations together sped up the algorithms by a factor of nearly 100. We repeated the experiments by varying the compositions of samples and parameter values *Length, Dist, Occur* with consistent results.

To evaluate the usefulness of our pattern discovery algorithm, we have incorporated it into a classifier (referred to as PA; Wang et al., 1994b, 1996; Chirn, 1996) to classify 698 groups of related proteins in the SWISS-PROT protein sequence databank version 27 (Bairoch and Boeckmann, 1992). These groups are categorized based on the documentation given in the PROSITE catalog version 11.0 (Bairoch, 1992). The groups comprise more than 15,000 proteins.

Our classifier works by taking 70% of the proteins in each group, using them as the training data, and finding active patterns from them. It then classifies the remaining 30% test sequences by matching them against the active patterns found for each group. The classifier gives the highest rank to the group with the most patterns matching with a test sequence.

For comparison purposes, we also ran the current best classifier, referred to as HH (Henikoff and Henikoff, 1991), on the dataset. HH generates a set of blocks for each group, where a block comprises ungapped aligned regions extracted from the sequences in the group. To classify a test sequence, HH matches the sequence against all the blocks and displays a collection of groups, ranked based on their relevance to the sequence.

Table 4.2 summarizes the classification results. A test sequence was classified correctly by HH (respectively, PA) if its group was ranked highest by HH (respectively, PA). The table also shows the results when the two

Table 4.2 Comparison between HH and PA.

Classification results	Percentage of the test sequences
HH was correct	93.5%
PA was correct	92.3%
HH and PA agreed and both were correct	86.7%
HH and PA agreed and both were wrong	0.5%
HH and PA disagreed and HH was correct	7.3%
HH and PA disagreed and PA was correct	4.8%
HH and PA disagreed and both were wrong	0.7%

classifiers agreed (i.e., the highest ranked group returned by both of them was the same) and disagreed on their rankings. Notice that the last five percentages in the table add up to 100%.

Thus, if HH and PA agree, the classification has a high likelihood of being correct. Specifically, the correct agreed upon classification divided by the total agreed upon classification is $86.7\%/(86.7\% + 0.5\%) = 99.4\%$. On the other hand, if HH and PA disagree, then the likelihood that one is right is $(7.3\% + 4.8\%)/(7.3\% + 4.8\% + 0.7\%) = 94.5\%$. Thus, the two classifiers give information that is complementary to one another. Using the two classifiers together, one can obtain high-confidence classifications (if they agree) or suggest a new hypothesis (if they disagree).

In addition to the protein classification, our pattern discovery algorithm also helps in DNA classification. The next section presents one such example.

4.2 Classification of DNA Sequences

DNA sequence classification is an important problem in computational biology (Gelfand, 1995). Given an unlabeled sequence S, a classifier makes predictions as to whether or not the sequence belongs to a particular class C. Many computer-assisted techniques have been proposed for constructing classifiers from a library of labeled sequences. In general, these techniques can be categorized into the following three classes:

- *Consensus search.* This approach takes a collection of sequences of the class C and generates a "consensus" sequence that is then used to identify sequences in uncharacterized DNA (Staden, 1984; Galas et al., 1985; Mulligan and McClure, 1986; Berg and von Hippel, 1987; Studnicka, 1987; O'Neill and Chiafari, 1989; Gelfand, 1995).
- *Inductive learning/neural networks.* This approach takes a set of sequences of the class C and a set of sequences not in C and then, based on these sequences and using learning techniques, derives a rule that determines whether the unlabeled sequence S belongs to C (Quin-

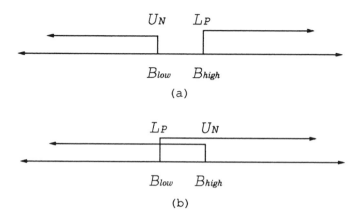

Figure 4.7 Illustration of the parameters B_{low} and B_{high} for cases $U_N \leq L_P$ (a) and $U_N > L_P$ (b).

queton and Moreau, 1985; Sallantin et al., 1985; Lukashin et al., 1989; Lapedes et al., 1990; Hirst and Sternberg, 1992; Gelfand, 1995; Loewenstern et al., 1995).

* *Sequence alignment.* This approach aligns the unlabeled sequence S with members of C using an existing tool such as Fasta (Lipman and Pearson, 1985; Pearson and Lipman, 1988) and assigns S to C if the best alignment score for S is sufficiently high.

We review here two new methods, based on our pattern discovery algorithm and a hash-based fingerprint technique, for calculating scores to classify DNA sequences. Our approach works by first randomly selecting a set B of sequences of the class C, referred to as "base data." Then we take another set of sequences of C, referred to as "positive training data," and calculate, for each positive training sequence, a score with respect to the base sequences. The minimum score thus obtained is called the positive lower bound, L_p. Next, we take a set of sequences not in C, referred to as "negative training data," and again calculate, for each negative training sequence, a score with respect to the base sequences. The maximum score thus obtained is called the negative upper bound, U_n. Let $B_{high} = \max \{L_p, U_n\}$ and $B_{low} = \min \{L_p, U_n\}$ (see figure 4.7). When classifying the unlabeled sequence S, we calculate S's score with respect to the base sequences, denoted c. If $c \geq B_{high}$, then S is classified as a member of C. If $c \leq B_{low}$, then S is classified as not a member of C. If $B_{low} < c < B_{high}$, then the "no opinion" verdict is given.

The two proposed classifiers differ in their ways of processing the base sequences and calculating scores for the training and unlabeled sequences. We describe each classifier in turn in the following subsections.

4.2.1 Pattern-Based Classifier

The first classifier, referred to as the *pattern-based classifier*, applies the pattern discovery algorithm presented in section 4.1.2 to find active patterns in the base data. Let $\mathcal{R}$ be a set of active patterns discovered from the set $\mathcal{B}$ of base sequences. Given a sequence S (which could be a training or an unlabeled sequence), the score between S and a pattern $P \in \mathcal{R}$, denoted $score(S, P)$, is defined as $|L|$ (i.e., the number of nucleotides in L), where L is the longest common substring of S and P. (The time complexity for finding the score is bounded by $O(|L|)$, and at worst $O(|S| \times |P|)$ [Clift et al., 1986; Cobbs, 1994].) The score of S with respect to the base sequences is defined as

$$score(S) = \max\{score(S, P) | P \in \mathcal{R}\} \times 100.$$

4.2.2 Fingerprint-Based Classifier

The second classifier, referred to as the *fingerprint-based classifier*, adopts a hash-based fingerprint technique to calculate scores (Califano and Rigoutsos, 1993; Wang et al., 1996). Let S be a sequence and let Seg be a segment, that is, a consecutive subsequence, of S. A *gapped fingerprint* f of Seg is a possibly noncontiguous subsequence of Seg that begins with the segment's first nucleotide. A gap at position p means that when forming f, we do not pick the nucleotide at position p in Seg. For example, let $S = $ ACGTTGCA. Then $Seg = $ ACGTTG is a segment of S and ATT is a fingerprint of Seg with two gaps (one gap at position 2 and one gap at position 3). The number of gaps allowed in these fingerprints is bounded by a parameter gap.

Building Fingerprint Files

Given the set $\mathcal{B}$ of base sequences, we pick the segments from each base sequence and hash each fingerprint of the segments into a file, as described below. Let S be a sequence in the base dataset $\mathcal{B}$. We take every segment Seg of length n from S and generate gapped fingerprints from Seg. The lengths of Seg's fingerprints range from 2 to $n-1$. Then we use a hash function h_k, $2 \leq k \leq n - 1$, to hash all fingerprints of length k to a fingerprint file $\mathcal{F}_k$. In the file, each fingerprint f is associated with a pair of integers (x, y). This pair serves as the position marker for f, where x indicates that f is generated from a segment of the xth sequence in $\mathcal{B}$ and y means that the first nucleotide of f occurs at the yth position in that sequence.

> **Example 4.2** Consider the following three base sequences: $S_1 = $ ACGTTGCA, $S_2 = $ ACCAGTG, $S_3 = $ CGGACTA. Suppose the length of seg-

Table 4.3   Gapped fingerprints (of lengths 2, 3, 4, 5, respectively) generated from the segment ACGTTG (the segment length is 6 and $gap = 2$).

	____	Fingerprints	____	____
	Two nucleotides	Three nucleotides	Four nucleotides	Five nucleotides
No gap	AC	ACG	ACGT	ACGTT
One gap	AG	ACT	ACGT	ACGTG
		AGT	ACTT	ACTTG
			AGTT	AGTTG
Two gaps	AT	ACT	ACGG	
		AGT	ACTG	
		ATT	AGTG	
			ATTG	

ments is 6. Then, for example, we obtain the following segments from S_1: ACGTTG, CGTTGC, and GTTGCA.

Now consider the segment Seg = ACGTTG taken from S_1. Suppose $gap = 2$. Then, we can generate the following three-nucleotide gapped fingerprints from Seg: ACG (no gap), AGT (one gap at position 2), ACT (one gap at position 3), ATT (one gap at position 2 and one gap at position 3), AGT (one gap at position 2 and one gap at position 4), and ACT (one gap at position 3 and one gap at position 4). Table 4.3 shows all gapped fingerprints generated from the segment ACGTTG.

Let f = XYZ be a fingerprint of length 3. Suppose the hash function h_3 is $h_3(f) = [num(X) \times 4^2 + num(Y) \times 4^1 + num(Z)] \bmod 7$, where $num(X)$ is X's ASCII value minus 64. Figure 4.8 shows the fingerprint file $\mathcal{F}_3$ for the three base sequences S_1, S_2, and S_3. Thus, for example, in bucket 1 in $\mathcal{F}_3$, GGA(3, 2) means that the fingerprint GGA is generated from S_3 and it starts from the second position in S_3.

Algorithm for Scoring

When calculating the score of a sequence S (whether it is a training or an unlabeled sequence), we segment S in the same way as for the base sequences and generate fingerprints from the resulting segments. We then hash the fingerprints, using the same hash functions as for the base sequences. When a match between S's fingerprint and a base sequence's fingerprint occurs, we add one to the score of an appropriate position in the base sequence, as illustrated below.

Consider again the base sequence S_1 = ACGTTGCA in example 4.2 and a sequence S = CGATGCAT. Consider the three-nucleotide fingerprint TGC

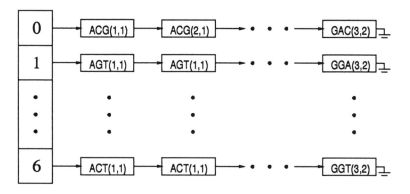

Figure 4.8   The fingerprint file $\mathcal{F}_3$ for the three base sequences in example 4.2. Note that some entries are duplicate [e.g., AGT(1, 1) in bucket 1]; they represent fingerprints with different numbers of gaps.

starting from the fifth position in S_1, and the same TGC starting from the fourth position in S. Let $q = 5$ and $p = 4$. We add one to the score of the position $q - p + 1 = 5 - 4 + 1 = 2$ in S_1. Intuitively, if we align the first nucleotide of S with the second nucleotide of S_1, we can see a match between the two corresponding fingerprints (figure 4.9). In general, if one aligns the first nucleotide of S with the kth position in S_1 that obtains n scores in total, one can see n matches of fingerprints in the alignment. Thus, aligning the first nucleotide of S with the position in S_1 with the highest score may yield the best alignment between the two sequences. This technique was pioneered by Califano and Rigoutsos (1993) for finding the best alignment between two DNA sequences.

Figure 4.10 summarizes our scoring algorithm. The result is a histogram of scores on the base sequences. Figure 4.11 illustrates an example. Here we are given the sequence $S = $ CGATGCAT and the three base sequences in example 4.2. The histogram is obtained after matching S's fingerprints with all the fingerprints of the base sequences.

Let B be a base sequence in $\mathcal{B}$ and let p be a position in B, $1 \leq p \leq |B|$. Let $score(B[p])$ represent the total scores added to the position p in B after applying the algorithm Scoring to the sequence S and the base sequences in $\mathcal{B}$. The score of B, denoted $score(B)$, is defined as

$$score(B) = \max\{score(B[p]) | 1 \leq p \leq |B|\}.$$

S_1: ACGTTGCA
             $|\ |\ |$
 S :   CGATGCAT

Figure 4.9   Illustration of the alignment between S_1 and S.

Input: A sequence S, a set $\mathcal{B}$ of base sequences and $\mathcal{B}$'s fingerprint files.
Output: A histogram of scores on the base sequences in $\mathcal{B}$.
/* Let $\mathcal{F}$ contain all fingerprints generated from S. */
for each fingerprint f in $\mathcal{F}$ **do**
   begin
      /* Let the length of f be k. */
      hash f using h_k and probe into the fingerprint file $\mathcal{F}_k$;
      for each match between f and a fingerprint $\hat{f}$ in $\mathcal{F}_k$ **do**
         begin
            /* Let the position marker associated with $\hat{f}$ be (i, q). */
            /*Suppose the first nucleotide of f occurs at the pth position in S.*/
            add one to the score of the position $q - p + 1$ in the ith base
            sequence in $\mathcal{B}$;
         end;
   end;

Figure 4.10   Algorithm Scoring.

The score of S with respect to the base sequences, denoted $score(S)$, is defined as

$$score(S) = \frac{\max\{score(B)|B \in \mathcal{B}\}}{|S|} \times 100.$$

4.2.3   Experiments and Results

The algorithms for the proposed classifiers were implemented in C programming language on a Sun SPARCstation 20 running the operating system

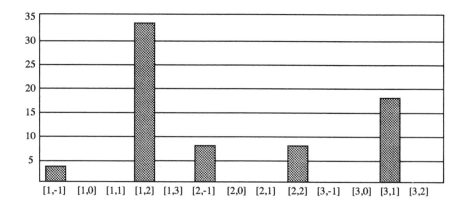

Figure 4.11   Histogram obtained after processing the sequence $S =$ CGATGCAT and the three base sequences in example 4.2. The y-axis shows scores. Each $[i, q]$ on the x-axis represents the qth position in the ith base sequence in example 4.2.

Table 4.4 Parameters and their base values used in classification experiments.

Parameter	Value	Description
$|\mathcal{B}|$	100	Number of base sequences (Alu)
$|\mathcal{T}_P|$	100	Number of positive training sequences (Alu)
$|\mathcal{T}_N|$	100	Number of negative training sequences (non-Alu)
NumTest	1,227	Number of unlabeled test sequences (Alu and non-Alu)
Length	11	Minimum length of active patterns used in the pattern-based classifier
Occur	15	Minimum occurrence number of active patterns used in the pattern-based classifier
Dist	0	Allowed distance between an active pattern and a base sequence used in the pattern-based classifier
n	8	Segment length used in the fingerprint-based classifier
gap	0	Number of gaps allowed in the fingerprint-based classifier

Solaris version 2.4. We compared the relative performance of the algorithms by using them to classify Alu sequences (Jurka et al., 1993; Claverie and Makalowski, 1994). Three hundred twenty-seven Alu sequences were selected from the database at the National Center for Biotechnology Information (NCBI; ftp at ncbi.nlm.nih.gov/pub/jmc/alu/ALU.327.dna.ref). Among them, 100 were used as base sequences, 100 were used as positive training sequences, and the other 127 were treated as unlabeled test sequences. The lengths of these sequences ranged from 76 to 379 base pairs (bp). In constructing non-Alu data, following the studies in Lukashin et al. (1989) and Demeler and Zhou (1991), we randomly generated 1,200 sequences that retained the correct nucleotide frequencies using the simulation programming package SIMSCRIPT (Kiviat et al., 1983). Among the data, 100 were used as negative training sequences and the other 1,100 were also treated as unlabeled test sequences. The lengths of these sequences ranged from 100 to 240 bp.

The pattern-based classifier found active patterns from the 100 base sequences. The active patterns had the form $*X*$ and had length ≥ 11, occurrence 15, and distance 0 (i.e., these patterns matched at least 15 base sequences without mutation). There were 556 active patterns, with lengths ranging from 11 to 22 bp. The fingerprint-based classifier fixed the segment length at 8 and *gap* at 0. Table 4.4 summarizes the parameters and their base values used in the experiments.

The metrics used to evaluate the effectiveness of our classification algorithms are precision rates (PR) and no-opinion rates (NR), where

$$PR = \frac{NumCorrect}{NumTest} \times 100\%$$

and

$$NR = \frac{NumNoOpinion}{NumTest} \times 100\%.$$

NumCorrect is the number of test sequences classified correctly, *NumNo-Opinion* is the number of test sequences obtaining the "no opinion" verdict, and *NumTest* is the total number of test sequences, 1,227 in our case. (A test sequence S in a class $\mathcal{C}$ is said to be classified correctly by an algorithm if S is determined to belong to $\mathcal{C}$ by that algorithm.)

Our experimental results showed that for the pattern-based classifier, $B_{high} = 1,100$, $B_{low} = 1,000$ and $PR = 99.5\%$, $NR = 0.0\%$. For the fingerprint-based classifier, $B_{high} = 68$, $B_{low} = 65$ and $PR = 99.4\%$, $NR = 0.08\%$. For comparison purposes, we also ran the Fasta classifier (Lipman and Pearson, 1985; Pearson and Lipman, 1988) on the same datasets. This classifier aligns a given unlabeled DNA sequence S with an Alu consensus taken from the database maintained in the Whitehead Institute for Biomedical Research. It classifies S as an Alu if the alignment score is greater than or equal to a preset threshold. (The score is the best score of S against both strands of the consensus, and the threshold used is 100.) Otherwise, S is classified as a non-Alu sequence. The tool does not give the "no opinion" verdict. The experimental results showed that Fasta achieved a 99.1% PR, lower than the PR of our classifiers.

We next conducted a series of experiments to examine the impact of the parameter values on the performance of the two proposed classifiers. To avoid the mutual influence of parameters, in each experiment we varied only one parameter's values, with the other parameters being fixed and having the values shown in table 4.4. It was found that both of the classifiers behave stably with varying $|\mathcal{B}|$, $|\mathcal{T}_P|$, and $|\mathcal{T}_N|$. The relative sizes of the base, positive training, and negative training datasets have little effect on the performance of the classifiers, provided that these sets are sufficiently large (e.g., with size ≥ 100). Figure 4.12 shows that the performance of the pattern-based classifier degrades as *Occur* becomes large. We found that the discovered patterns in Alu sequences generally have low occurrence numbers. When $Occur > 35$, very few patterns were discovered and thus they cannot well characterize the sequences. Likewise, using short active patterns (e.g., with *Length* $= 5$) and large distance values (e.g., *Dist* $= 3$) yields poor performance, since these patterns may appear, by chance, in both Alu and non-Alu sequences. No trend is evident regarding n and *gap* used in the fingerprint-based classifier. However, programs using a small *gap* (e.g., *gap* $= 0$) run much faster than programs using a large *gap* (e.g., *gap* $= 4$).

In sum, to achieve good performance, one should use sufficiently many (e.g., 100) sequences as base, positive training, and negative training data when constructing the classifiers. For the pattern-based classifier, using long active patterns (e.g., with *Length* $= 11$) with a small distance value (e.g., *Dist* ≤ 1) is good. Such patterns appear quite uniquely in Alu sequences and therefore characterize the sequences well. For the fingerprint-based

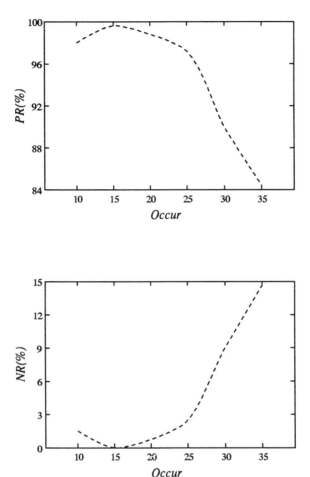

Figure 4.12 *PR* and *NR* as a function of *Occur* used in the pattern-based classifier.

classifier, using segments of length 8 without gaps is good, as it requires less running time and space while achieving an acceptably high precision.

4.3 Generalizations and Future Work

In this chapter we have presented some techniques for pattern discovery and classification in biosequences. Experimental results on both protein and synthetic data showed that our discovery algorithm and optimization heuristics work effectively. Our empirical study also showed a promising result when applying the proposed classification algorithms to Alu sequences.

Our current work has three main goals:

1. By incorporating a consensus approach into the algorithms presented here, we are developing new techniques to classify more subtle sequences such as promoters and splice junctions. This work will eventually contribute to gene recognition in DNA.
2. The current classification tools return results without stating the confidence level. We plan to study the statistical behavior of the classifiers so as to enhance the quality of their output.
3. We want to extend our pattern discovery techniques to high-dimensional data, such as 3D proteins, and develop new techniques for protein classification and clustering based on the discovered patterns. In this direction we have recently obtained some preliminary results (Wang et al., 1997).

The software developed from the work is available from the authors. Readers interested in obtaining the software can send a written request to jason@cis.njit.edu or shasha@cs.nyu.edu or visit our Web site at `http://www.cis.njit.edu/~discdb` for details.

Further Information

A good survey to DNA sequence classification is presented in Gelfand (1995). Hunter (1993) provides a sampling of artificial intelligence approaches to problems of biological significance.

The *Proceedings of the International Conference on Computational Molecular Biology* published by the Association for Computing Machinery (ACM) document the latest development in the field of biological computing. The *Proceedings of the International Conference on Scientific and Statistical Database Management* published by the Institute of Electrical and Electronics Engineers (IEEE) Computer Society contain articles on biomolecular data management.

The quarterly journal *Journal of Computational Biology* reports advances in biological computing. For subscription information, contact Mary Ann Liebert, Inc., publishers, 1651 Third Avenue, New York, NY 10128; phone 1-800-M-LIEBERT, fax (212) 289-3347.

Part II. Finding Patterns in 3D Structures

Chapter 5

Motif Discovery in Protein Structure Databases

Janice Glasgow, Evan Steeg, and Suzanne Fortier

The field of *knowledge discovery* is concerned with the theory and processes involved in the representation and extraction of patterns or motifs from large databases. Discovered patterns can be used to group data into meaningful classes, to summarize data, or to reveal deviant entries. Motifs stored in a database can be brought to bear on difficult instances of structure prediction or determination from X-ray crystallography or nuclear magnetic resonance (NMR) experiments. Automated discovery techniques are central to understanding and analyzing the rapidly expanding repositories of protein sequence and structure data.

This chapter deals with the discovery of protein structure motifs. A *motif* is an abstraction over a set of recurring patterns observed in a dataset; it captures the essential features shared by a set of similar or related objects. In many domains, such as computer vision and speech recognition, there exist special regularities that permit such motif abstraction. In the protein science domain, the regularities derive from evolutionary and biophysical constraints on amino acid sequences and structures. The identification of a known pattern in a new protein sequence or structure permits the immediate retrieval and application of knowledge obtained from the analysis of other proteins. The discovery and manipulation of motifs—in DNA, RNA, and protein sequences and structures—is thus an important component of computational molecular biology and genome informatics. In particular, identifying protein structure classifications at varying levels of abstraction allows us to organize and increase our understanding of the rapidly growing protein structure datasets. Discovered motifs are also useful for improving the efficiency and effectiveness of X-ray crystallographic studies of proteins, for drug design, for understanding protein evolution, and ultimately for predicting the structure of proteins from sequence data.

Motifs may be designed by hand, based on expert knowledge. For example, the Chou-Fasman protein secondary structure prediction program (Chou and Fasman, 1978), which dominated the field for many years, depended on the recognition of predefined, user-encoded sequence motifs for α-helices and β-sheets. Several hundred sequence motifs have been cataloged in PROSITE (Bairoch, 1992); the identification of one of these motifs in a novel protein often allows for immediate function interpretation. In recent years there has been much interest and research in automated motif discovery. Such work builds on ideas from machine learning, artificial neural networks, and statistical modeling and forms the basis for the methods for protein motif discovery described in this chapter.

The search for protein structure motifs begins with the knowledge that some proteins with low sequence similarity fold into remarkably similar 3D conformations. Moreover, even globally different structures may share similar or identical substructures (Unger et al., 1989; Rooman et al., 1990), as predicted by Pauling et al. (1951) and verified in many of the early crystallographic experiments. Researchers have discovered and cataloged important motifs at the secondary, supersecondary, and tertiary structure levels (Taylor and Thornton, 1984; Ponder and Richards, 1987). Recently, there has also been interest in the automated discovery and use of structural motifs at levels finer than that of secondary structure (Unger et al., 1989; Rooman et al., 1990; Hunter and States, 1991).

In this chapter we describe several types of protein motifs and present approaches to knowledge discovery and their application to the problem of structure motif discovery. We conclude the chapter with a presentation of criteria by which a motif, and associated motif discovery techniques, may be judged.

5.1 Protein Motifs

The study of relations between protein tertiary structure and amino acid sequence is a topic of tremendous importance in molecular biology. The automated discovery of recurrent patterns of structure and sequence is an essential part of this investigation. These patterns, known as protein motifs, are abstractions over one or more fragments drawn from proteins of known sequence and tertiary structure. The Protein Data Bank (Bernstein et al., 1977) is the main source of data regarding the 3D structure of proteins.

Protein motifs can be roughly divided into four categories (as illustrated in figure 5.1): *sequence motifs*, linear strings of amino acid residues (or residue classes) with an implicit topological ordering; *sequence-structure motifs*, sequence motifs that associate residues in the motif with secondary structure identifications; *structure motifs*, 3D objects that correspond to portions of a protein backbone, possibly combined with side chains; and *structure-sequence motifs*, structure motifs in which nodes of the graph are annotated with sequence information. The distinction between sequence

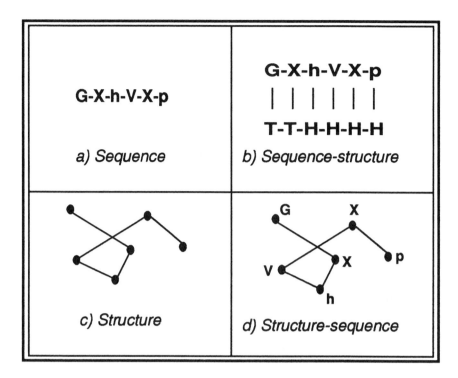

Figure 5.1 Various types of protein motifs. X, any residue; G, glycine; V, valine; H, α-helix; T, turn; p, polar; h, hydrophobic.

and structure motifs has previously been considered by Thornton and Gardner (1989) and Conklin (1995a).

Different types of motifs have different purposes. For example, protein sequence motifs can facilitate the incremental acquisition of sequence data into knowledge bases organized according to sequence similarity (Taylor, 1986b). Protein structure motifs can be used as building blocks for protein model building in crystallography (Jones and Thirup, 1986). Finally, protein structure-sequence motifs are useful in automated structure prediction, model building, and model design (Unger et al., 1989). They are also applicable to automated approaches to protein structure determination from crystallographic data (Fortier et al., 1993; Glasgow et al., 1993).

Below we discuss the different types of protein motifs in more detail and present recent studies that involved the discovery of such motifs.

5.1.1 Sequence and Sequence-Structure Motifs

Protein sequence motifs are the most commonly encountered motif type in the molecular biology literature. It is generally assumed that similarities in protein sequence indicate structural and functional similarity. Thus, the discovery of sequence motifs from structurally similar proteins or protein fragments is an important method for uncovering relationships between structure and sequence. Protein sequence motifs can facilitate the incremental acquisition and indexing of sequence data into knowledge bases organized according to sequence similarity (Taylor, 1986b).

Finding motifs in sequences involves two main steps: (1) assembling a training set of sequences with common structure or function and then (2) analyzing the training set for regions of conserved amino acids. Sequence motifs may be discovered from a maximal alignment of one or more protein sequences, followed by the abstraction of residues at aligned positions. Conserved residues are those identical at corresponding alignment positions. In recent years, the use of hidden Markov models has become predominant in the sequence modeling area (Krogh et al., 1994). There is an extensive literature on the comparison of sequence motifs.

Much of the work on protein structure prediction is based on the a priori definition of sequence motifs that predict a certain type of secondary structure identifier. These sequence-structure motifs (referred to by Thornton and Gardner, 1989, as "structure-related sequence motifs") have an inherent directionality of implication from sequence to structure. The work of Rooman and Wodak (1988) associates each amino acid in a motif with a standard secondary structure identifier (e.g., motif b in figure 5.1). This work demonstrated that, while associations sufficient to predict a complete protein structure were not derived, a number of reliably predictive motifs do exist.

5.1.2   Structure and Structure-Sequence Motifs

Structure motifs constitute the building blocks that can be used to describe the tertiary structure of a protein. The accurate prediction of protein tertiary structure from amino acid sequence, while theoretically possible, remains one of the great open problems in molecular biology. One way of addressing this has been to break the problem into two parts: predicting secondary structure from sequence, followed by the problem of packing the secondary structure into 3D conformations.

Although secondary structures are the most commonly considered structure motifs, there has been increased interest in the automated discovery and use of structural patterns at both a finer and coarser level than that of α-helix or β-sheet. Unger et al. (1989), for example, report the discovery of motifs for structural fragments containing six amino acids, whereas Taylor and Thornton (1984) consider motifs at the level of supersecondary structure.

Global tertiary structure motifs typically characterize protein families and superfamilies (Orengo et al., 1994), though several global motifs (such

as the so-called TIM-barrel) are shared by proteins that are not evolutionarily related or functionally similar.

The abundance of sequence information produced by large-scale sequencing efforts, combined with the increasing rate of structure information
coming from X-ray crystallography and NMR, is producing the kind of huge
and diverse databases necessary for both model-based and "memory-based"
structure prediction and determination, as well as for phylogenetic analysis in evolutionary biology. Concomitant with the increasing availability of
these different data types is a growing belief that protein structure prediction and recognition methods ought to be, like the protein folding process
itself, a simultaneously global and local, bottom-up and top-down application of constraints. For these reasons, the last several years have seen
an increasing focus on integrated, multiple-resolution, multiple-view protein databases. This general trend and the associated opportunities to use
genomic, structural, functional, and evolutionary information to reinforce
one another and fill in one another's gaps have led researchers to look for
structural motifs that can be associated with and predicted from sequence
motifs.

A few researchers have added sequence annotation, or computed sequence-structure correlation, after completion of their structure motif discovery.
Unger et al. (1989) used a k-nearest-neighbors method to find clusters in
the space of protein backbone fragments of length 6 residues. They tabu-
lated the frequencies of amino acids types at every position, producing a
sequence motif for each of their approximately 100 structure motifs. Their
results indicated that the local 3D structure of a fragment can sometimes
be predicted by the assignment of the fragment to a motif based on these
frequency tables. Rooman et al. (1990) produced a physicochemical properties motif for each of the 4–10 structural classes they discovered in different
runs of a hierarchical clustering of fragments of lengths 6–10. Zhang and
Waltz (1993) used a variant of k-means clustering on structure fragments
of size 7 and then tested the χ^2 significance of the association of their 23
local structure classes with particular amino acid combinations.

In contrast, other researchers have attempted to produce structure-
sequence motifs, or just highly predictable structure motifs, by discovering motifs in sequence space and structure space concurrently. Lapedes et
al. (1995) simultaneously trained two neural networks, one taking local
sequence fragments as input and the other taking the corresponding local
tertiary structure fragments, using an objective function that maximized
the correlation between the two networks' outputs. The effort described
in their article, representing preliminary steps in a larger, ongoing project,
succeeded in finding novel secondary structure classes that are more predictable from amino acid sequence than the standard helix, sheet, and coil
classes.

Conklin (1995b) used conceptual clustering to produce mutually predictable sequence and structure motifs and, by treating both sequence and
structure fragments within the same model-theoretical framework, avoided

the "confused dual semantics" displayed by many other attempts to relate sequence and structure.

As an alternative to the traditional approach of predicting structure from sequence, *inverse protein folding* involves predicting sequence from structure. For example, one can use genetic algorithms or other search methods to generate and test many possible sequences to see which ones might fold into a given structural pattern. Protein threading algorithms are used to answer the question of whether and how each such sequence "fits" onto a structure motif characterized by particular physicochemical environments and attributes.

5.1.3   Hierarchical Motifs

A number of groups have worked on the discovery of motifs at two or more levels of protein organization, simultaneously or in stages.

The work of Zhang and Waltz (1993) is a good example of this idea. Their motif discovery process proceeded in three stages. First, they selected a set of objectively definable primitives that compactly carried the local structural geometry of protein backbone segments. Second, the database of local structure segments, represented in terms of these new canonical feature vectors, was clustered using a k-means method. The researchers found that 23 such novel secondary structure classes produced the best fit to the data. Third and finally, they represented longer stretches (corresponding roughly to supersecondary structure) of protein backbone, in terms of sequences of symbols corresponding to the local structure classes found in step 2, and designed and trained a finite-state machine to recognize the sequences. Thus, the description of the trained finite-state machine represented a set of de novo supersecondary structure motifs discovered on top of a set of de novo secondary structure motifs. Building on this work, Zhang and colleagues later tested other input representations and clustering parameters to arrive at a set of structural building blocks that, they claim, have generality across large and diverse sets of proteins.

Conklin (1995a) introduced a category of hierarchical protein motifs that captures, in addition to global structure, the nested structure of the subcomponents of the motif. This representation was used to discover recurrent amino acid motifs, which were then used for the expression of higher level protein motifs.

5.2   Computational Approaches to Knowledge Discovery

Knowledge discovery has been defined as "the non-trivial process of identifying valid, novel, potentially useful, and ultimately understandable patterns in data" (Fayyad et al., 1996b). Generally, the automated discovery process is interactive and iterative. Brachman and Anand (1996) divide this process into several steps: developing an understanding of the application domain, creating a target dataset, data cleaning and preprocessing, finding useful features with which to represent the data, data mining to search for patterns of interest, and interpreting and consolidating discovered patterns.

Several approaches to knowledge discovery have been proposed and applied. Below we present an overview of some of these approaches derived from research in machine learning and statistics. We also discuss their applications to the problem of protein structure and structure-sequence motif discovery.

5.2.1 Clustering Approaches

Clustering is a discovery task that seeks to identify a finite set of classes or clusters that describe a dataset. Appropriately derived clusters have predictive and explanatory power and lead to a better understanding of a complex dataset. Derived clusters (or classes) may be mutually exclusive or overlapping. Clustering techniques may generally be divided into three categories, numerical, statistical, and conceptual, described in more detail below.

Cutting across all three major categories are other aspects by which to distinguish clustering techniques. For example, both *agglomerative* and *divisive* clustering techniques exist. Agglomerative techniques use a starting point consisting of as many clusters as instances; in divisive techniques the starting point consists of a single cluster. Clustering techniques can also be differentiated on the basis of whether they allow for overlapping clusters or whether they only produce disjoint partitions. A clustering technique may be *incremental* or *nonincremental* depending on whether all of the observations are available at the outset of a clustering exercise. Another important distinction is whether an approach incorporates flat or hierarchical representations. Hierarchical methods have a special appeal in the protein science domain, in which both the structural organization of a protein and the evolutionary process that generated it can be described hierarchically.

Numerical Clustering

In numerical clustering, samples are viewed from a geometric perspective as a set of data points in an n-dimensional space, where n is the number of attributes used to characterize each data point. The goal of the clustering exercise is to partition the data points, grouping similar points together. Figure 5.2 illustrates a simple clustering of a banking dataset into three

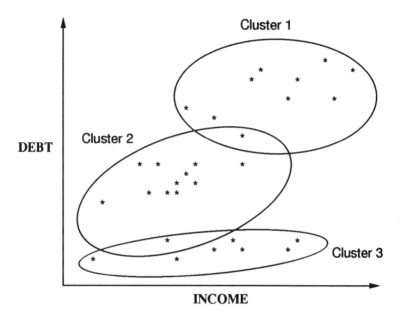

Figure 5.2 Clusters derived for a set of data relating income to debt to determine the viability of giving a loan.

overlapping classes. Such classes could be used to predict good/bad risk groups for loans.

Distance metrics are used to measure similarity/dissimilarity among data points, while criterion functions help measure the quality of the data partition. Thus, numerical clustering techniques generally rely on quantitative attributes. In structural biochemistry, attributes such as interatomic and interresidue distances and bond and torsion angles are often considered in performing numerical clustering. The objective in most numerical clustering methods is to maximize some combination of intraclass similarity and interclass distance.

Rooman et al. (1990) use an agglomerative numerical clustering technique to discover structure motifs, which are represented by prototypical fragments. In this method, only C_α positions are used in the description of the amino acid position. Similarity between fragments is measured using a root-mean-square (RMS) metric of the inter-C_α distances. A fragment is an instance of a motif class by virtue of being within an RMS distance threshold from the prototypical motif for that class.

Unger et al. (1989) report an experiment in structure-sequence motif discovery. Hexamers, described by C_α positions of residues, are clustered using a k-nearest-neighbor algorithm. As in Rooman et al. (1990), similarity between hexamers is measured according to a distance metric. In this approach, structures are first aligned using a best molecular fit routine, and absolute coordinates—rather than intramotif distances—are compared.

Bayesian Statistical Clustering

Although statistical clustering techniques typically use numerical representations and metrics, the history and philosophy behind them mandate a separate treatment. In standard numerical methods such as single-linkage, total-linkage, and mutual nearest-neighbor clustering (Jarvis and Patrick, 1973; Willett, 1987b), the goal is to assign data points to clusters. In contrast, the goal of a Bayesian latent class analysis, for example, is the construction of a plausible generative model of the data, that is, to explain the data.

In a latent class analysis approach to finding class structure in a set of data points, one begins with an underlying parameterized model. For example, one might posit that a set of points represented by a 2D scatterplot was generated by a 2D Gaussian (normal) distribution. Or, the data might be better explained by a mixture, or weighted sum, of several Gaussian distributions, each with its own 2D mean vector and covariance matrix. In this approach, one tries to find an optimal set of parameter values for the representation of each data point. Optimality may be defined in terms of maximum likelihood, Bayes optimality, minimum description length (MDL), or minimum message length (MML). The Bayes, MDL, and MML formalisms, in particular, put the somewhat vague objectives of standard numerical clustering—a trade-off between intraclass and interclass distances—onto firmer theoretical ground, forcing practitioners to make all assumptions and biases explicit and quantifiable. An MDL objective function, for example, imposes a cost on the total length of the shortest description of the data. This cost includes the cost of specifying the model itself, of specifying the class parameters, and of specifying the residual difference between the true positions of the data points versus the positions predicted from class parameters. Roughly, the model and class parameter costs can be called *complexity*, and the residual differences cost can be called *distortion* (Buhmann and Kuhnel, 1993). The complexity/distortion trade-off is clear: more, smaller classes raise complexity but lower distortion (because each data point is closer to its class center). The general term "Bayesian" is often informally applied to most or all of these mathematically similar and philosophically related methodologies (McLachlan and Basford, 1988; Baxter and Oliver, 1994).

Philosophical foundations aside, the major practical differences distinguishing this class of statistical methods from most other numerical clustering methods include the following:

- Classes may overlap in a probabilistic sense; a single point may typically be shared among two or more classes.

- The same objective function can be used to measure and guide both the particular class centers, variances, and memberships, and the total number of classes.
- The clustering process induces a generative model of the data, meaning the model can be used both to classify new points and to generate datasets of realistic-looking points (hence, in this case, plausible protein substructures).

Just as Bayesian and related statistical methods have gained prominence in sequence analysis, through Gibbs sampling (Lawrence and Reilly, 1990) and hidden Markov models (Krogh et al., 1994), so have they come to prominence in structure motif discovery.

Hunter and States (1991) used AutoClass, a general-purpose Bayesian classification tool (Cheeseman et al., 1988), to discover structure motifs for amino acid segments of length 5. Motifs are represented in this work by a probability distribution over Cartesian coordinates for the backbone atoms of each residue. They discovered 27 classes of structural motifs (in the highest likelihood classification), where the majority of coordinates were assigned with a high probability to a single class.

Another general purpose statistical classification and modeling tool, Snob (Wallace and Dowe, 1994), has also been used to discover novel local structure motifs (Dowe et al., 1996). In contrast to previous work, the researchers in this project recognized the superiority of circular von Mises distributions, as opposed to Gaussians, in modeling angular data such as backbone dihedrals. The Snob program searches for MML-optimal classifications of objects that may be defined in terms of any number of real number, angular, or discrete attributes, each of which is modeled with the appropriate type of probability distribution.

The mutually supervised sequence and structure networks designed by Lapedes et al. (1995) can be recast within a Bayesian unsupervised learning framework. Indeed, other neural network learning methods, such as those used in predicting aspects of protein structure (McGregor et al., 1989; Kneller et al., 1990), can also be understood within and often improved by a Bayesian analysis (MacKay, 1992).

Conceptual Clustering

Conceptual clustering techniques share with their numerical counterparts the goal of partitioning the data into natural groupings. They have, however, an additional goal, which is to characterize the clusters in terms of simple and meaningful concepts, rather than in terms of a set of statistics. These methods predominantly use qualitative attributes. Some of these commonly considered in protein motif discovery are proximity and spatial configuration.

The term *concept formation* is normally used to refer to incremental conceptual clustering algorithms. Following the definition of Gennari et al. (1989), concept formation can be described as follows:

given a sequential presentation of objects and their associated description, find (1) clusters that group these objects into classes, (2) a summary description (i.e., a concept) for each class, and (3) a hierarchical organization for these concepts.

Several useful concept formation algorithms currently exist, including UNIMEM (Lebowitz, 1987) and Cobweb (Fisher, 1987). These systems rely, however, on an object representation being expressed as a list of attribute-value pairs. This representation is not the most suitable for the domain of protein structure, where the most salient features of the object may involve relationships among its parts. An emerging area of interest in machine learning is the design of structured concept formation algorithms in which structure objects are formed and then organized in a knowledge base.

IMEM is a concept formation method specifically designed for objects or scenes described in terms of their parts and the interrelationships among these parts (Conklin and Glasgow, 1992; Conklin et al., 1996). These relationships may be topological (e.g., connectivity, proximity, nestedness) or spatial (e.g., direction, relative location, symmetry). A molecular structure is represented in IMEM as an image, which comprises a set of parts with their 3D coordinates, and a set of relations that are preserved for the image. The IMEM algorithm uses an incremental, divisive approach to build a subsumption hierarchy that summarizes and classifies a dataset. This algorithm relies on a measure of similarity among molecular images that is defined in terms of their largest common subimages.

The IMEM approach has been implemented as a system to perform conceptual clustering with protein structure data (Conklin et al., 1996). Figure 5.3 illustrates a classification exercise for a given protein fragment. Assuming the initial hierarchy of figure 5.3(a), the fragment in figure 5.3(b) is initially stored as a child of the most specific subsuming motif. The result of this step is depicted in figure 5.3(c). A concept formation step then occurs where a novel motif, which subsumes both the new fragment and a previously classified fragment, is generated. This last step is illustrated in figure 5.3(d).

The Cobweb system (Fisher, 1987) performs incremental, unsupervised concept formation using an information-theoretic evaluation function to construct a concept hierarchy. Specialized versions of Cobweb and Auto-Class (Cheeseman et al., 1988) have been used to classify pairs of secondary structure motifs in terms of supersecondary motifs (Schulze-Kremer and King, 1992). The results of applying these clustering methods were combined to form a consensus clustering.

Conceptual clustering has also been applied to sequence motif discovery. Shimozono et al. (1992) developed a decision tree method to classify membrane protein sequences according to functional classes using a variation of the conceptual learning system ID3 (Quinlan, 1986).

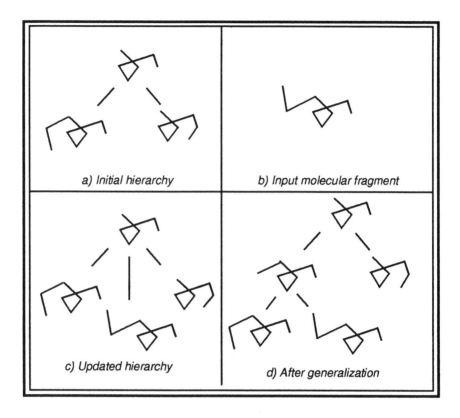

Figure 5.3   Development of a concept hierarchy for molecular images.

5.2.2   Nonclustering Techniques

There are a number of techniques that are not based on clustering per se but that also discover, or support the discovery of, protein structure and structure-sequence motifs. Prominent among these are (1) methods based on graph matching that find recurrent motifs in terms of maximal common substructure or similar combinatorial criteria, (2) feature-selection methods that may be viewed as performing an essential first step in motif discovery, and (3) methods for finding empirical folding potentials.

Koch et al. (1996) used a variation of maximum clique enumeration to find maximal common subtopologies in a set of protein structures. They represented protein secondary structures as unlabeled directed graphs and restricted the usual clique enumeration algorithm to only those cliques that represent connected substructures.

It is important to ensure that the right information is input into a motif discovery process if one expects meaningful output. Feature discovery and feature selection are often crucial initial steps within a larger motif discovery task. In the protein structure analysis domain, many primitive features are selected by hand, based on prior expert knowledge. Other features may be discovered as special subsets or combinations of hand-picked features. Principal component analysis is commonly employed for this task on vector representations, because it can take an n-dimensional representation of a dataset and return a reduced representation in terms of $m < n$ decorrelated dimensions. This has advantages in later stages of machine learning or other processing. Other, analogous methods may be used on nonvector representations. For example, algorithms exist for finding pairs (Korber et al., 1993; Klingler and Brutlag, 1994) or k-tuples (Steeg, 1997) of correlated residue positions in sets of aligned protein sequences. Such correlations may bespeak evolutionarily conserved structural or functional interactions and form the basis for a particularly useful kind of structure-sequence motif. The discovery of correlations and associations in protein representations need not be limited to residues alone; one group of researchers has used a general-purpose data mining tool (Agrawal et al., 1996) to find substructures that correlate with particular functional features (Satou et al., 1997).

Another important type of protein motif is the empirical folding potential. Sippl (1990) has put this expanding enterprise on firm theoretical ground by using an inverse Boltzmann equation to translate between theoretical force field terms and empirical database frequencies. By performing statistical studies of amino-amino proximity relationships, core/surface preferences of particular amino acids, and so on, Sippl and others (Jones et al., 1992) have built motifs into objective functions that can supplement or replace theoretical potentials in structure prediction and threading.

5.3 Assessing Protein Motifs and Discovery Methods

A reading of the relevant research literature in protein analysis suggests at least six broad criteria by which to measure protein motifs and, by extension, the methods used to define and discover them:

1. *Predictability* is the degree to which a motif representing one level or facet of protein structure or function may be predicted from knowledge of another. For the local structure motifs designated as "secondary structure," predictability is the ability to accurately predict secondary structure classes from amino acid sequence.
2. *Predictive utility* is the flip side of the predictability criterion. For example, if one takes the view of secondary structure as an intermediate-level encoding between primary structure (sequence) and tertiary structure, then predictive utility ought to be some measure of the gain in accuracy in predicting tertiary structure with a particular encoding, as compared with prediction using other possible encodings.

Another more direct measure might be the degree to which a particular set of proposed motifs, corresponding to secondary structure classes, constrain the ϕ and ψ angles of the included structure fragments.

3. *Intelligibility* refers to the ease with which researchers and practitioners of protein science can understand a given structure motif and can incorporate its information into their own work. Many factors affect intelligibility. For example, a discovered structure class that contains one-third traditional α-helix, one-third traditional β-sheet, and one-third coil is harder to explain than one that correlates almost perfectly with α-helix. Also, for example, a motif expressed in first-order logic with terms for well-known biochemical aspects such as amino acid names and dihedral bond angles is easier to understand than a motif represented only in a set of several hundred neural network connection weight values. Further aspects of motif intelligibility are discussed below.

4. *Naturalness*, or the equally unwieldy word "intrinsicness," means the degree to which a motif captures some essential biochemical or evolutionary properties or some essential class structure in the space of protein sequence or structure fragments under consideration. Some clustering methods are infamous for finding ersatz clusters in uniformly distributed data. Other clustering methods produce results very dependent on their starting point. To avoid such results it is important to carefully choose appropriate representations and attributes for classification.

5. *Ease of discovery* refers to the computational complexity and data complexity of the methods required to discover the motif.

6. *Systematicity* is the degree to which a motif discovery method is derived from explicitly stated principles and the degree to which the method can repeatably be applied to diverse data and produce consistent results.

5.4   Issues in Protein Motif Discovery

What kinds of issues determine the desirability of particular motifs and methods under the criteria listed above? We discuss below some important considerations for carrying out a motif discovery exercise.

5.4.1   Use of Both Sequence and Structure Data

The discovery of sequence motifs, along with its associated goal of multiple sequence alignment, has been the mainstay of computational molecular biology from the beginning. A central hypothesis is that similarity of sequence implies similarity of structure and function, and evolutionary conservation of sequence implies conservation of structure and function. Thus, it is difficult in a sense to find any work on sequence motifs that does not involve sequence-structure motifs. Much of the vast literature on consensus sequences, sequence profiles, and hidden Markov modeling of protein families describes the discovery of motifs over sets of sequences that are already known to correspond to one particular overall tertiary structure. But where researchers have sought to define family-independent subsequence motifs that carry important structural information, they have not necessarily succeeded. In a study described in Rooman et al. (1990), 11 out of a set of 12 sequence-structure motifs claimed to be predictive of secondary structure were found not to be.

There are two major limitations to the above methods. First, the sequence information is incorporated into the structure motifs after the latter have been defined. This approach is not designed to, and is not likely to, produce structure classes predictable from amino acid sequence. Second, as noted by Conklin (1995b), in the methods outlined above, each structure motif may be associated with only a single sequence motif; there is no provision made for associating a structure motif with a disjunction of different sequence motifs (except in the narrow disjunction implicit in the abstraction of several very similar sequences into a more abstract motif).

The IMEM method proposed by Conklin, Fortier, and Glasgow addresses both of these limitations by representing both sequence and structure objects in the symbolic format of a spatial description logic, a restricted first-order logic used to describe and manipulate concepts. Motif discovery in this system occurs through similarity-based clustering (structured concept formation) of combined sequence-structure representations. The sequence-structure predictability is built into the discovery process and enforced through a series of tests that measure the strength and significance of associations between sequence and structure motifs. For example, the ratio M_+/N—where N is the number of protein fragments assigned to the sequence portion of the motif and M_+ is the number of fragments assigned to the joint structure-sequence motif—must be greater than 0.8. This ensures that more than 80% of all the instances of the sequence motif have the same structure, and therefore that the sequence may predict the corresponding structure. Another test is a χ^2 test applied to a 2×2 contingency table for each structure-sequence motif. This test assesses the significance of association, over all protein fragments in the dataset, between the sets of fragments assigned or not assigned to the sequence portion of the motif and the sets of fragments assigned or not assigned to the structure portion.

5.4.2 Number of Classes, or Motifs

The number of different motifs, or classes, sought in a discovery procedure has important impact on both the information-theoretic aspects of predictive utility of the resulting motifs and the general intelligibility and usefulness of the motifs to molecular biologists.

First, in terms of the distortion versus complexity trade-off in clustering and latent class modeling, it is clear that more classes generally imply lower distortion and higher complexity costs. That is, the larger the number of classes and hence class centroids (exemplars, control points), the closer a given point will be to the centroid of its own class, ceteris paribus. But the larger the number of classes, the more bits it takes to encode each data point in terms of its class-label encoding.

Though the MDL communications paradigm (minimizing a total number of bits transmitted between a hypothesized sender and receiver) is a somewhat artificial theoretical tool, it does reveal important practical aspects of data models. As the number of classes in a model of structure fragment data increases, a real trade-off becomes apparent. Each motif becomes more specific, in that it carries more detailed local structural information about a smaller set of fragments. On the one hand, this might make subsequent tertiary structure prediction easier, because structure-packing considerations are made more precise. On the other hand, there is a loss of abstraction, a greater number of parameters to optimize in the motif discovery algorithm, and a potentially greater difficulty in finding statistically significant estimates of frequencies and probabilities of motifs and features.

A growing consensus in computational molecular biology favors classes less coarse than the standard two to five secondary structure classes. Conklin (1995b) cites three reasons:

1. First, he claims that there exist wide discrepancies between different methods of assigning secondary structure designations from crystallographically determined structures. This point is debatable. It appears to other observers that the Kabsch and Sander (1983) standard is both well founded and widely accepted. However, to the extent that discrepancies do exist, one must take care that a prediction system is not just modeling the idiosyncrasies of particular structure definition rules.

2. A great number of fragment patterns tossed into the large default class "random coil" are neither random nor undefinable. Add to this the fact that different kinds of helices, and different kinds of β-strand configurations, can be observed, and there is a case to be made for additional subclasses of the three major classes.

3. Secondary structure packing analysis is a nontrivial task, and more accurate descriptions of local backbone structure—as should result from motif discovery with larger numbers of classes—can make the task much easier.

5.4.3 Locality: Size of Input Fragments

The size of sequence and structure fragments input to motif discovery systems is another issue closely related to the question of abstraction versus specificity. Smaller fragments imply smaller, more localized motifs. On the one hand, smaller motifs mean that a greater number of them are needed to represent an entire sequence or structure, and hence a greater number of parameters are used in later stages of a modeling or prediction task. On the other hand, smaller motifs also correspond to more frequently occurring patterns, and therefore problems in probability estimation are minimized.

One must also carefully consider domain-specific and goal-specific criteria when choosing fragment size: over what lengths of sequence and of structural backbone chain do the phenomena of interest manifest themselves? For example, individual β-strands can be captured with fragments of size 6–12, typically, but what about the turns between strands? How much information about the strand is conveyed by the nearby turns, and vice versa? How much information do different strands carry about each other? How much nonlocal information is necessary to determine a sequence fragment's propensity to "become" a helix or a strand or a stretch of coil, for example?

The information-theoretic and the biophysical issues here are deep and complex. An empirical, trial-and-error approach might be reasonable in attacking this problem. For example, a fragment size (prediction "window" size) of 13 was found to be effective in previous work on secondary structure prediction (Qian and Sejnowski, 1988; Kneller et al., 1990). In such studies it was found that smaller windows failed to provide sufficient local contextual information for prediction of the secondary structure of the central residue in the window; for windows of length > 13, the marginal gains in extra contextual information were swamped by noise.

Another issue is fixed- versus variable-length motifs. For reasons described earlier in this chapter, most of the reported projects in structure-sequence motif discovery looked for short, fixed-length motifs of size 5–8 (Unger et al., 1989; Hunter and States, 1991; Zhang and Waltz, 1993). It has been observed, however, that different phenomena manifest themselves over different lengths (such as helices and turns), and different pieces of a protein have evolved and are conserved over different lengths of sequence. Interactions between distant residues, and the failing of structure prediction methods to take them into account, is one of the hypothesized reasons for the limited prediction success that has been achieved. In general, with longer motifs, more contiguous residues can be predicted, and less tertiary alignment of predicted portions needs to be performed. Thus, it is too restrictive to discover motifs of only one size. An advantage of the IMEM approach (Conklin et al., 1996) is that it can discover variable-length motifs.

5.4.4   Representation

Perhaps the most important initial choice to be made in designing a motif-discovery method is what kind of representation to use for motifs. The differences between some of the options are huge, as large as the traditional gulf between the "symbolic/logical" and "numeric/statistical" camps in artificial intelligence research. The stakes can also be high, both in terms of the amount of interesting information captured by the resulting motifs and in terms of our ability to understand and communicate the information.

For structure classification, numerical clustering methods dominate the field (Rooman et al., 1990; Hunter and States, 1991; Zhang and Waltz, 1993). There are good reasons for this. First, structures are geometric and physical objects, and the representation of such objects—in terms of vectors, angles, and chemical properties—is an old and strong tradition in the physical and computational sciences. Second, the use of numeric features and statistical clustering techniques is very amenable to the use of well-defined objective functions, thus enabling a generally principled approach and the use of well-understood optimization procedures.

If the goal is intelligibility of derived motifs, there is no contest—logical representations are preferred. Clearly it is difficult to look at a set of hundreds of connection weights or means and variances and see anything resembling a motif. However, once a set of classes has been discovered, there is no major obstacle to finding more recognizable and descriptive motifs after the fact. The set of sequences, for example, corresponding to a particular structural class can be aligned, clustered, and so on, using standard methods, and consensus sequences can be produced.

A virtue of the IMEM method for representations (Conklin, 1995a; Conklin et al., 1996) is that sequence and structure motifs are represented using a common formal syntax. Unlike the other structure-sequence motifs mentioned in this chapter and surveyed in Conklin (1995b), the structure-sequence motifs of IMEM do not inherit a "confused dual semantics," where a sequence is interpreted using one formalism and a structure another. The knowledge representation formalism implemented in IMEM presumably enables it to be integrated more easily into larger, multilevel, multiview protein analysis systems wherein many different kinds of features are used to predict other features.

Although a logic representation may be preferable, it is often difficult to determine the appropriate primitive concepts and qualitative relationships necessary for conceptual clustering. An integrated approach that incorporates both numeric and conceptual methods could address this issue. As a first step, numeric techniques could be used to perform an initial classification and derive parameters to be applied in a second step that would use conceptual clustering to derive meaningful (logical) concepts for the discovered motifs.

5.4.5   Intrinsic versus Extrinsic Clustering Criteria

Implicit in some of the above discussion is a concept of intrinsic versus extrinsic criteria for clusters and motif discovery. In the multistage process and multiple levels of description that characterize protein structure prediction, as in machine vision and speech recognition, there is a tension between the "best" intermediate representation language suggested by optimizing local, "current-level" criteria (what are the best clusters in $\Phi\Psi$ space?) and those suggested by optimizing "next-level-up" criteria (which clusterings produce classes that work well as primitive symbols in a tertiary structure encoding?). This is a fundamental issue not yet addressed sufficiently in the computational molecular biology domain. There may be insights to be gained by examination of other domains that require motif discovery at several levels of organization, such as computer vision, speech recognition, and natural language understanding.

5.5   Conclusion

The investigation of relations between protein tertiary structure and amino acid sequence is of enormous importance in molecular biology. The automated discovery of recurrent patterns of structure and sequence is an essential component of this investigation. This chapter has provided an overview of existing methods for protein structure motif discovery and some of the outstanding issues involved in this field.

Traditional machine learning and knowledge discovery techniques are not always appropriate for learning in the protein structure domain. This is mainly because they assume that similarity between objects is measured as a distance function based on simple attributes and values. The representation issues for structure motifs, however, are more complex; similarity is often judged in terms of spatial relations among parts of a structure as well as in terms of attributes of the structure and its parts. Another distinction is the size of existing datasets and the implied efficiency considerations resulting from the vastness and complexity of the data. Thus, there is an ongoing challenge to find appropriate methods for gathering information and knowledge from the ever-growing repositories of protein data and, in particular, for understanding the intricate relationship between sequence and structure data.

Further Information

Recent research in the area of knowledge discovery is surveyed in Fayyad et al. (1996a). A good overview of the area of sequence motif discovery can be found in Lathrop et al. (1993). The doctoral dissertations of Steeg (1997) and Conklin (1995a) include more complete reviews of statistical and conceptual clustering techniques and their application to protein structure motif discovery.

The *Proceedings of the International Conference on Intelligent Systems for Molecular Biology*, published by the American Association for Artificial Intelligence (AAAI) and MIT Press, and the *Proceedings of the Pacific Symposium on Biocomputing*, published by World Scientific, contain articles related to protein motif discovery. The *Proceedings of the National Conference on Artificial Intelligence* and the *Proceedings of the International Conference on Knowledge Discovery and Data Mining*, both published by AAAI Press, contain general articles on machine learning and motif discovery. A special issue of the journal *Machine Learning* (vol. 21, 1995) focuses on learning and discovery techniques for molecular biology.

Chapter 6

Systematic Detection of Protein Structural Motifs

Kentaro Tomii and Minoru Kanehisa

It is widely believed that the prediction of the 3D structure of a protein from its amino acid sequence is important because the structure will help understand the function. As the number of protein structures resolved is increasing, most predictive methods have become based on the knowledge of the repertoire of 3D folds taken by actual proteins. We must emphasize, however, that this type of structure prediction, or fold recognition, concerns the overall folding of the polypeptide chain. Since two similar folds could be due to entirely different sequences and even two similar sequences could have different functions, it is unlikely that successful fold recognition will uncover any functional clue that cannot otherwise be obtained by sequence analysis alone.

In contrast to the global feature of 3D folds, the concept of structural motifs or local structures is far more important in understanding protein function. It has been revealed that there are common local folding patterns that appear in many proteins of globally different structures and that are involved in conserved function. In addition, the local sequence patterns associated with these local structures are also often conserved, though the whole sequences can be quite different. At the supersecondary structure level there are, for example, $\beta\alpha\beta$-unit, EF hand, and helix-turn-helix motifs (figure 6.1). Various dehydrogenases have a common structural motif called Rossman fold, which is composed of two consecutive $\beta\alpha\beta$-units, and most of those proteins also have the sequence motif GxGxxG around the nucleotide binding region (Wierenga and Hol, 1983; Wierenga et al., 1986).

The EF hand consisting of the helix-loop-helix structure (Tufty and Kretsinger, 1975) occurs in many calcium-binding domains, and the residues that participate in ligand binding are well conserved. The helix-turn-helix

Name	βαβ-unit	EF-hand	Helix-turn-helix (CRP family)
Function	nucleotide binding	calcium binding	DNA binding
Sequence	[LIVM]-[AG]-[LIVMT]-[LIVMFY]-[AG]-x-G-[NHKRQSC]-[LIVM]-G-x(13,14)-[LIVMT]-x(2)-[FYCTH]-[DNSTK]	D-x-[DNS]-[ILVFYW]-[DENSTG]-[DNQGHRK]-[GP]-[LIVMC]-[DENQSTAGC]-x(2)-[DE]-[LIVMFYW]	[LIVM]-[STAG]-[RHNW]-x(2)-[LI]-[GA]-x-[LIVMFYA]-[LIVSC]-[GA]-x-[STAC]-x(2)-[MST]-x-[GSTN]-R-x-[LIVMF]-x(2)-[LIVMF]
Structure			

Figure 6.1 Examples of functional motif in proteins that are known to involve common sequence patterns, common structural patterns, and conserved functions (see table 6.3 for more examples).

motif that involves about 20 residues appears in a class of DNA-binding domains, and glycine tends to be conserved at a special position in the turn whose conformation corresponds to the left-handed helix.

A number of known sequence motifs are registered in the motif libraries such as PROSITE (Bairoch and Bucher, 1994) that compile the relationships between sequence patterns and functions. In contrast, the organization of known structural motifs is not well developed, especially in relation to the function involved. If we can detect systematically as many structural motifs as possible, and clarify the correlation, if any, with the corresponding amino acid sequence motifs, we should then be able to predict both the protein local structure and the function at the same time from its amino acid sequence. The 3D structural information in this case will give an additional functional clue of how protein molecules might interact with other molecules based on a stereochemical point of view.

The size of the Brookhaven Protein Data Bank protein structure database (Bernstein et al., 1977) is continuously and rapidly growing, and it has become too expensive computationally to compare protein structures exhaustively in the 3D coordinates for extraction of structural motifs. In the field of sequence analysis, the techniques of dynamic programming (Needleman and Wunsch, 1970; Smith and Waterman, 1981b; Goad and Kanehisa, 1982) are considered most time-consuming, but actually they are rapid enough to make exhaustive comparison of all sequences (Gonnet et al., 1992) in the sequence database, which contains two orders of magnitude more proteins than the structural database. To exploit the efficiency of sequence analysis, we have adopted the strategy to represent the 3D structure of a protein in terms of a symbol string of structural states (Matsuo and Kanehisa, 1993). The suite of sequence analysis techniques can then be utilized for automatic extraction of both structural motifs and sequence motifs.

Our procedure to convert protein coordinate datasets into symbol strings is analogous to the quantization procedure that is widely used in the field of digital signal processing. The quantization procedure converts continuous signals such as sound and image into digital ones adequate for computer-based manipulation. Similarly, our procedure reduces continuous data in the conformational space of short polypeptide segments to discrete categories, each of which is assigned a symbol. The conformation of the entire polypeptide chain is represented by the collection of conformations taken by overlapping segments, that is, by a symbol string. This can also be compared with the DSSP program (Kabsch and Sander, 1983) that defines the secondary structure state for each amino acid residue of a protein, thus representing the protein conformation by a symbol string. While DSSP strings contain a number of blanks for the nonassigned coil state that is actually a collection of widely different conformations, our algorithm assigns a different symbol for a different state according to its conformation, regardless of its secondary structure.

In this chapter, we first give a brief review of different attempts to define a set of conformational states taken by short polypeptide segments, which may be considered as building blocks of protein structures, to provide the basis of representing the protein structure as a sequence of discrete states. We then summarize the method and results of our previous work (Matsuo and Kanehisa, 1993; Tomii and Kanehisa, 1996), followed by discussion and future directions.

6.1   Structural Building Blocks

Procedures to classify short polypeptide segments, say, four to nine residue long, into groups of similar conformations have been developed (Unger et al., 1989; Rackovsky, 1990; Rooman et al., 1990; Prestrelski et al., 1992; Matsuo and Kanehisa, 1993; Unger and Sussman, 1993; Oldfield and Hubbard, 1994; Schuchhardt et al., 1996; Fetrow et al., 1997) where different clustering methods have been applied to a high-dimensional space of protein segment conformations as summarized in table 6.1. The resulting clusters or representatives are sometimes called "structural building blocks." In the classical secondary structure definition based on the template method, such as the DSSP program, considerable portions of the protein chain do not fit in the templates, which leave them as random coil regions. In contrast, our classification methods overcome the limitation of the secondary structure definition by categorizing the whole conformational space that short polypeptide segments can adopt in proteins.

Table 6.1 The list of clustering methods used for classifying short peptide segments in proteins.

No.	Distance measure	Np^a	SL^b	Clustering method	Nc^c	Applications	Reference
1	Distances and pseudotorsion angles of C_α atoms	123	4	Indexing of the "Generalized Bond" matrix	4	Classification of proteins	Rackovsky, 1990
2	Distance of C_α atoms/inertia coefficient	75	4–7	Hierarchical clustering (plus Ramachandran plot)	≥4	Identification of structural motifs	Rooman et al., 1990
3	Distances and pseudotorsion angles of C_α atoms	116	7	K-means clustering algorithm	6	Identification of structural motifs	Fetrow et al., 1997
4	The pseudoangles and pseudotorsion angle of C_α atoms	83	4	Analysis of contour plots and Ramachandran plots	22	Model building	Oldfield and Hubbard, 1994
5	RMS distance after superposition	93	7	See text	37	Structure alignment	Matsuo and Kanehisa, 1993
6	RMS distance after superposition	82	6	Modified k-nearest-neighbor clustering algorithm (further clustering by threshold)	81	Structure verification	Unger and Sussman, 1993
7	ϕ and ϕ torsion angles	136	9	Kohonen algorithm	100	Structure comparison	Schuchhardt et al., 1996
8	RMS distance after superposition	4	6	Modified k-nearest-neighbor clustering algorithm (further clustering by threshold)	103	Structure verification	Unger et al., 1989
9	Distances and pseudotorsion angles of C_α atoms	14	8	Cost function tuning	113	Calibration of FTIR measurements	Prestrelski et al., 1992

a. The number of proteins in the dataset.   b. The length of the peptide segment.   c. The number of generated clusters.

To derive the representative set of oligopeptide conformations, different authors employed different datasets and different measures for defining the similarity or distance between segment conformations and between clusters of segment conformations. As shown in table 6.1, the number of proteins used in constructing the building blocks has been increasing along with the growth of the Protein Data Bank.

The distance measures between a pair of segments were mainly based either on the root mean square (RMS) distance between the corresponding C_α atoms after the best superposition or on the RMS deviation of the corresponding interatomic distances of C_α atoms (distance maps) with or without the α-carbon torsion angles. Only Schuchhardt et al. (1996) used the ϕ and φ backbone torsion angles for their classification, which had been known to be sensitive to rotations in the peptide plane (Karpen et al., 1989). For the definition of the distance between two clusters, most of these methods employed the conformational data directly or their linear transformations. However, Fetrow et al. (1997) used a nonlinear transformation, that is, the output values of the trained neural network, which effectively reduced the dimensionality of the input conformational data.

As the result of the studies in table 6.1, common features have emerged. There are three levels of clustering resolutions, which we call low, medium, and high resolutions containing roughly 4, 30, and 100 clusters, respectively. The most prominent condensation of the cluster(s) corresponds to α-helix elements, while β-strands have been recognized for their variability at the high-resolution clustering. For example, Unger and Sussman (1993) showed that there were three classes for the extended hexamers whose secondary structure assignment by the DSSP program was EEEEEE ("E" corresponds to the strand in a β-ladder). Other common observations include the correlation between the contents of the local structure types and the global architecture of proteins (Rackovsky, 1990; Schuchhardt et al., 1996) and the amino acid preferences at certain positions within different categories of peptide segments (Fetrow et al., 1997).

What is the optimal number of groups and the segment length for such a classification? It may appear that if a longer segment length is chosen it becomes possible to detect more detailed similarities and differences. However, whether the longer segment length actually results in an increased number of groups depends on the clustering method. In fact, there is not a clear and unvarying tendency in table 6.1. For example, it seems natural intuitively that the length of four residues is suitable to treat classical β-turns, and the higher level classification of seven residues long could identify $\alpha\beta$- and $\beta\alpha$-loop motifs (Rooman et al., 1990). However, even using the broader criterion of categorization into only six groups from seven residue segments, Fetrow et al. (1997) succeeded in recognizing the detailed features of helix caps (ends) and strand caps by observing a specific series of category assignments along the protein chain. Thus, we consider that the types of structural patterns detected depend on both of the two parameters, the number of clusters, and the segment length.

The number of clusters is also related to the homogeneity, that is, the degree of similarity between segment conformations within each cluster. In fact, all except methods 3 and 7 in table 6.1 assigned a given threshold of similarity that then determined the resulting number of clusters, while methods 3 and 7 employed an unsupervised algorithm for clustering given a preassigned number of clusters. It may appear objective to use an unsupervised algorithm, but it is unclear whether the number of clusters preassigned is optimal. Fetrow et al. (1997) examined various numbers of clusters, from 2 to 12, and found that the value 6 was most appropriate.

6.2 Symbol String Representation of 3D Structures

The appropriate level of clustering should depend on the purpose of quantizing the conformational space. Matsuo and Kanehisa (1993) aimed to extract structural motifs by systematic comparison of structures represented in one-dimensional symbol strings. They thus needed a manageable-sized set of conformational categories that would be assigned different symbols. Here we show their algorithm to select representatives of the conformational groups among all segments of proteins in the dataset (see table 1 of Matsuo and Kanehisa, 1993).

The distance between two segments is defined by the RMS distance between the corresponding α-carbon atoms after the best superposition (Kabsch, 1976, 1978). The algorithm has two parameters l and d_s; l is the segment length, which was set at 7 residues, and d_s is the sampling interval, which was set at 2.01 Å. The concept of sampling interval (Matsuo and Kanehisa, 1993; Crippen and Maiorov, 1995) is schematically shown in figure 6.2. The parameter values were manually selected to generate a manageable number of representatives and to satisfy an appropriate level of homogeneity so that the members within the same category were similar with one another.

The algorithm of Matsuo and Kanehisa for selecting representatives can be described in the pseudocode style shown in figure 6.3. This algorithm is not deterministic. That is to say, the set of representatives depends on the order of choosing the segments. However, the set of representatives always satisfies the following properties: (i) the distance between any pair of representatives is always greater than the sampling interval distance, and (ii) the minimum distance from each segment to one of the representatives is always equal to or less than the sampling interval distance. Therefore, when the set of representatives has been determined, each segment in the dataset can then be assigned to a category approximated by a representative with an error less than the sampling interval distance.

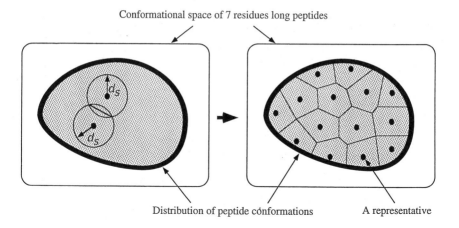

Conformational space of 7 residues long peptides

Distribution of peptide conformations   A representative

Figure 6.2   A schematic drawing of quantizing peptide segment conformations, where d_s denotes the sampling interval.

Matsuo and Kanehisa converted the coordinate dataset of 93 proteins into symbol strings; the resulting set of categories (representatives) is shown in table 6.2. The third column of the table lists the most frequent DSSP secondary structures; blank (coil) assignment is represented here by a period. Note that the seven-residue segments in 93 proteins were given to the algorithm as they appear in the database, which resulted in this particular

```
SELECTING_REPRESENTATIVES (d_s, S, R)
/* Initially, the set S contains all the segments of all the proteins, */
/* and the set of representatives R is empty. */
choose arbitrarily a segment s from the set S;
add s to the set R, and delete it from S;
while S is not empty do
    begin
        choose arbitrarily a segment u from S;
        dist(u) ← ∞;
        for each representative r of R do
            begin
                calculate the distance between u and r, d(u, r);
                if d(u, r) < dist(u) then dist(u) ← d(u, r);
            end;
        if dist(u) > d_s then add u to R;
        delete u from S;
    end;
```

Figure 6.3   Algorithm of Matsuo and Kanehisa (1993) for selecting representatives, in pseudocode style.

Table 6.2 The 37 groups of similar peptide conformations among the 93 proteins.

Group symbol	Number of members (%)	Secondary structure
N	2531 (16.52)	HHHHHH
#	1979 (12.92)	EEEEEE
C	1005 (6.56)	HHHHHH
X	868 (5.67)	.S.EEEE
Z	775 (5.06)	EEEE.ST
V	747 (4.88)	T...EEE
@	740 (4.83)	EEE.S..
5	511 (3.34)	..HHHHH
W	458 (2.99)	ET.....
H	415 (2.71)	EEE.TT.
A	389 (2.54)	.TT...E
D	385 (2.51)	E..HHHH
E	375 (2.45)	...TT..
U	341 (2.23)	..SS...
2	339 (2.21)	H.TT...
1	302 (1.97)	HHHHHT.
B	271 (1.77)	.TT..EE
G	269 (1.76)	E..SSS.
T	229 (1.50)	ETS..EE
O	202 (1.32)	HTTS..E
Y	197 (1.29)	..S..S.
6	194 (1.27)	HHHTT..
L	184 (1.20)	HHTT.EE
R	182 (1.19)	TTS.TT.
M	180 (1.18)	HHHTTT.
9	159 (1.04)	.SSST..
Q	157 (1.03)	T...S..
F	151 (0.99)	EE.TTEE
8	133 (0.87)	..TT..E
7	126 (0.82)	E.TT..E
S	115 (0.75)	.T..TTH
K	107 (0.70)	E.TTS.E
4	100 (0.65)	..SHHHH
3	90 (0.59)	E.TTT..
I	87 (0.57)	EE.TTS.
P	14 (0.09)	SSSSSS.
J	13 (0.09)	HHTTTS.
Total	15320	

set of representatives. The mean "within-category" distance was 1.28 ± 0.30 Å, which should be compared with the sampling interval distance of 2.01 Å. As shown in table 6.2, the most populated categories were, not surprisingly, those consisting of helices and β-strands.

6.3 Detection of Structural Motifs by Structure Comparisons

Once the coordinate data are converted into symbol strings, two protein structures can be compared by sequence alignment algorithms. Matsuo and Kanehisa used the modified Goad-Kanehisa algorithm (Goad and Kanehisa, 1982), which is a generalization of the Needleman-Wunsch algorithm, to find significant local similarities of folding patterns. As a consequence of the exhaustive comparison of the dataset, they obtained 858 significant similarity pairs that satisfied certain criteria. They detected, among others, the nucleotide-binding motif of the Rossman fold and the calcium-binding motif of the EF hand structure, both of which showed good correlations of the structural pattern, the sequence pattern, and the functional implication. We show here the case of the Rossman fold.

The three pairs, {4ADH(192-236), 6LDH(20-65)}, {4ADH(192-236), 1-ABP(35-78)}, and {1GD1O(1-47), 6LDH(20-65)}, contained the $\beta\alpha\beta$-unit followed by an α-helix at its carboxy-terminus. The RMS distances were, respectively, 2.02, 2.78, and 3.14 Å. Except for 1ABP(35-78), the structures were nucleotide-binding regions of dehydrogenases. Figure 6.4 shows the multiple alignments of these local structures in the symbol representation, together with the corresponding secondary structure assignments by the DSSP program, and the amino acid sequences. For each site of the multiple alignment of the structural motif, the minimum category that matched the amino acid substitution patterns was consistent with the 10 categories of amino acids according to Taylor (1986a; see table 3 of Matsuo and Kanehisa, 1993). The amino acid sequence motif obtained was generally in agreement with the fingerprint (Wierenga and Hol, 1983; Wierenga et al., 1986), although there were some differences caused largely by the presence of arabinose-binding protein (1ABP), which was functionally different from dehydrogenases.

Table 6.3 shows some examples of structural motifs that are known to be functionally important and that also contain conserved sequence patterns (shown with PROSITE accessions). These motifs are usually first identified at the sequence level, followed by the identification of structural patterns. It remains to be seen how many of the structural patterns detected by the methods such as Matsuo and Kanehisa's can be correlated to the sequence patterns stored in PROSITE. There have also been reports on certain structural motifs implicated in biological function despite the lack of sequence conservation. For example, dinucleotide-binding loops, mononucleotide-binding loops, and loops of the flavodoxin motif have been identified by scanning the Protein Data Bank with 12-residue segments (Swindells, 1993). These motifs are all superimposed to the probe structures within the RMS deviation 1.0 Å.

Table 6.3 Examples of structural motifs with functional relevance.

Name	Structural pattern	Sequence pattern (PROSITE accessions)	Biological function
βαβ-unit (Rossman fold)	βαβ	PS00065, etc.	Nucleotide binding
EF hand	Helix-loop-helix	PS00018	Calcium binding
Helix-turn-helix	Helix-turn-helix	PS00042, PS00519, etc.	DNA binding
Homeobox domain	Helix-turn-helix	PS00027	DNA binding
Leucine zipper	2 α-helices	PS00029	DNA binding
SH2 domain	2 α-helices and 6 or 7 β-strands	PS50001	Phosphotyrosine recognition
SH3 domain	5 or 6 β-strands	PS50002	Proline-rich peptide binding
PH domain	2 β-sheets and α-helix	PS50003	G protein binding
C2 domain	β-sandwich	PS00499, PS50004	Phospholipid binding
Cyclic nucleotide binding domain	3 α-helices and β-barrel	PS00888, PS00889	cAMP or cGMP binding

(a) Main chain conformations

```
                      +          +          +          +
1ABP ( 35- 78):  ##ZW-GD5CNNNCCCCNN16L-GWVXV#@@-WR9YCCCNNN
6LDH ( 20- 65):  ##ZWVWF45NNNNNNNNN168BAUVX##Z@-WGF5NNNNNN
4ADH (192-236):  ##ZWVWF45NNCCCCCNN16AW-@VX##Z@-WG95CCCNNN
1GD1O(  1- 47):  ##ZWVWF45NNNNCNNCN2A92OXZ@VXZ@GURD5NNNNNN
```

(b) Secondary structures

```
                         +          +          +
1ABP ( 35- 78):  EEE  -SHHHHHHHHHHHHHT -     B  -S SS TTHHHHHH
6LDH ( 20- 65):SSEEEEE  SHHHHHHHHHHHTTT  SEEEEE -S HHHHHHHHHHH
4ADH (192-236):T EEEE  SHHHHHHHHHHHHHS -  EEEE -S GGGHHHHHHHT
1GD1O(  1- 47): EEEEEE  SHHHHHHHHHHTT SSEEEEEEE SS HHHHHHHHHEE
```

(c) Amino acid sequences

```
                      +          +          +          +
1ABP ( 35- 78):VIKIAVP-DGEKTLNAIDSLAASG-AKGFVICT-PDPKLGSAIVAKA
6LDH ( 20- 65):YNKITVVGVGAVGMACAISILMKDLADEVALVD-VMEDKLKGEMMDL
4ADH (192-236):GSTCAVFGLGGVGLSVIMGCKAAGA-ARIIGVD-INKDKFAKAKEVG
1GD1O(  1- 47):AVKVGINGFGRIGRNVFRAALKNPDIEVVAVNDLTDANTLAHLLKYD
                      * * * * * *
```

Figure 6.4   The multiple alignments of βαβ-containing structures in dehydrogenases according to (a) the symbol representation of seven-residue segment conformations along the polypeptide chain, (b) the secondary structures assigned to each residue by the DSSP program, and (c) the corresponding amino acid sequences. The consensus sequence pattern GxGxxG for nucleotide binding is shown by asterisks (redrawn from Matsuo and Kanehisa, 1993).

6.4 Detection of Structural Motifs by Sequence Comparisons

The strategy described above converts coordinate data into symbol strings, applies sequence comparison techniques, and identifies protein structural motifs. Because the amount of sequence information far exceeds the amount of 3D structural information in the current databases, it is desirable to detect structural motifs from the analysis at amino acid sequence data. This is a structure prediction problem, but in this case the prediction involves local structures rather than the global structures of polypeptide chain folds.

One of the simplest approaches in that direction is to develop an alignment procedure that reflects structurally equivalent or similar regions. Then, a functionally relevant structural motif may be detected from the multiple alignment of functionally related sequences. In these attempts it is critical to construct a similarity scoring matrix, also called a mutation (substitution) matrix, for amino acid sequence comparisons to detect structural similarity. In addition to the standard matrices (Dayhoff et al., 1978; Henikoff and Henikoff, 1992) used for sequence similarity, which were compiled from the amino acid sequence data, there have been reports on the matrices compiled from the protein structural data. For example, the matrices were made by taking into account the distribution of backbone torsion angles (Niefind and Schomburg, 1991; Kolaskar and Kulkarni-Kale, 1992). Structural comparison methods were used to count amino acid replacements for deriving the matrices (Risler et al., 1988; Johnson and Overington, 1993). Another matrix was based on the spatial preferences of amino acids that reflected residue contacts in the protein tertiary structures (Qu et al., 1993).

The effectiveness of these matrices was shown in the original articles, but this can be examined from a different viewpoint. We collected 42 published matrices for amino acid similarity measures, based on both sequence data and structural data, and examined the correlations among them (Tomii and Kanehisa, 1996). This was done by cluster analysis, where the distance between each pair of matrices was defined by the correlation coefficient calculated from the 210 $[20 + (20 \times 19/2)]$ elements of an amino acid mutation matrix. The result of the cluster analysis is represented by the minimum spanning tree shown in figure 6.5. Thus, the matrices obtained from structural data, which are hatched nodes in figure 6.5, indicate different tendencies from the matrices obtained from sequence data, except for the one by Johnson and Overington (1993). It is therefore expected that the alignments made by these mutation matrices can be different and can be useful for predicting structural features of proteins (Kolaskar and Kulkarni-Kale, 1992).

The above approach of using a structurally derived mutation matrix can be extended to include additional features. For example, because the amino acid substitution patterns in proteins are not uniform in different structure classes, such as inside α-helices and outside β-strands, the use of separate

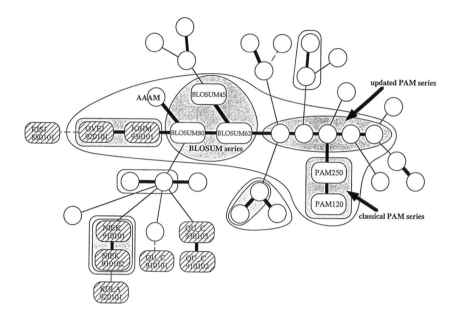

Figure 6.5   The minimum spanning tree of showing the similarity of the 42 amino acid mutation matrices (redrawn from Tomii and Kanehisa, 1996). Each node corresponds to a matrix. The shaded areas and the outer contours correspond to the clusters identified by single-linkage clustering with the threshold correlation coefficient of 0.96 and 0.92, respectively. The thick lines denote the clusters identified by complete-linkage clustering with the threshold correlation coefficient of 0.82. The hatched nodes correspond to the structure-based matrices: RISJ (Risler et al., 1988), OVEJ (Overington et al., 1992), JOHM (Johnson and Overington, 1993), NIEK (Niefind and Schomburg, 1991), KOLA (Kolaskar and Kulkarni-Kale, 1992), and QU_C (Qu et al., 1993).

mutation matrices for different structural classes should be more efficient for aligning structurally similar regions (Lüthy et al., 1991; Overington et al., 1992). The substitution patterns for the classes of amino acids (e.g., Taylor, 1986a) can be used for the prediction of structural features such as solvent accessibility and secondary structures for each site of multiply aligned protein sequences (Thompson and Goldstein, 1996). Conceptually, these extensions of the mutation matrix are similar to the 3D profile used in the threading, which takes into account the environmental factor that each amino acid exists in the 3D structures of actual proteins (Tomii and Kanehisa, 1996).

6.5   Discussion and Future Directions

The strategy of converting the 3D conformations of a short polypeptide chain segment into discrete states is, conceptually, an extension of assigning discrete secondary structure states to each residue of a polypeptide chain. Because the strategy would reflect only local structures determined by the segment length, it has not enjoyed wide acceptance among mainstream protein structure analysis, which is more concerned with the global folding of the entire polypeptide chain. From a biological point of view, however, it is the local structures that are most closely related to biological functions. The symbol string representation of protein structures is suitable for detecting such local structural motifs that often correlate with the amino acid sequence motifs. As exemplified in figure 6.1, it will be of great practical use to compile a dictionary of all the known local patterns of function-sequence-structure relationships.

The major advantage of assigning discrete states to local structures is, of course, the resulting one-dimensional symbol representation of the protein structure, which makes it possible to perform protein structure comparisons rapidly and systematically by using sequence comparison techniques. Earlier work by Usha and Murthy (1986) treated the pseudotorsion and pseudobond angles of consecutive C_α atoms along the protein chain as a symbol string. Recently, Laiter et al. (1995) divided the O(i-1)-C(i-1)-C(i)-O(i) pseudotorsion angle into 18 categories (symbols), constructed a scoring matrix for comparing symbol strings, and found that the OCCO angle was correlated with the ϕ, φ backbone dihedral angles or the protein secondary structures. In contrast to their conversion method, Matsuo and Kanehisa's (1993) procedure seems more sensitive to the tertiary structure of proteins.

Toward understanding the function-sequence-structure relationship of proteins, we reviewed in this chapter only the approach of starting from the 3D coordinates of protein structures and identifying structural motifs. The complementary approach is to start from the amino acid sequence data and to identify sequence motifs that can be correlated with common structural patterns and conserved functions. There have been a number of articles, including two of ours (Ogiwara et al., 1992, 1996), that report systematic methods to extract sequence motifs and to correlate with 3D structural features. Recently an interesting method has been proposed by Han and Baker (1996), who first identified recurring local sequence motifs by clustering substitution classes of amino acids at each site of multiply aligned protein sequences, and then analyzed the correlation between the sequence motifs and local structure patterns (see their figure 1).

Would the systematic analysis of protein 3D structural data and amino acid sequence data be sufficient for understanding the function-sequence-structure relationship of proteins? We must emphasize, for example, that the function of a DNA-binding protein will not be fully understood until the function of the target DNA is also understood. In other words, the analysis of single molecules will not be sufficient; the analysis of "interactions" among molecules is bound to be critical for ultimately understanding the protein function. We consider the major problem here to be the lack of an appropriate database and have thus initiated the efforts to computerize all known molecular interactions and molecular pathways in living cells under the project name KEGG (http://www.genome.ad.jp/kegg/). Just as the sequence databases and the Protein Data Bank initiated the current activity of computational biology, we hope KEGG will open a new possibility in this area.

Further Information

A more detailed review of the several classification methods of short protein segments is presented in Unger (1994). A review on the protein structure comparison methods utilizing dynamic programming techniques and other widely used methods that have not been covered here is presented in Orengo (1992).

Acknowledgments

This work was supported in part by a Grant-in-Aid for Scientific Research in the priority area "Genome Science" from the Ministry of Education, Science, Sports, and Culture of Japan. The computation time was provided by the Supercomputer Laboratory, Institute for Chemical Research, Kyoto University.

Chapter 7

Representation and Matching of Small Flexible Molecules in Large Databases of 3D Molecular Information

Isidore Rigoutsos, Daniel Platt, Andrea Califano, and David Silverman

In recent years, the need to process and mine available information repositories has been increasing. The variety of the data contained in the targeted databases has given rise to a variety of tools for mining them, and computers have assumed an increasingly important role in this process.

One of the many domains in which this scenario has been repeated is that of the drug discovery and design process. Computers have helped researchers to quickly eliminate unlikely drug candidates, to home in on promising ones, and to shorten the lead-compound-search cycle.

Researchers are helped in this multidisciplinary effort by accessing proprietary and public resources containing crystallography, nuclear magnetic resonance, toxicology, pharmacology, and other types of data. Using the computer to filter out unlikely candidates can greatly shorten the length of a cycle in this iterative process.

Some scenarios encountered in the context of the drug design process include

(a) a pharmacophore model that has been proposed from several active molecules—one wishes to determine other molecules that either corroborate or refute the model;

(b) a set of untested molecules that exhibit biological activity—one wishes to identify relationships between their 3D structure and the activity;

(c) a ligand that has been proposed to be active in a certain conformation--other molecules that mimic the ligand's behavior are sought.

The common element in all of these cases is that they are in essence searches for member elements in one or more repositories, each of the elements having some desired properties or behavior.

Let us take a step back and reexamine the problem we are trying to solve. Two basic elements of the problem are "representation" and "storage." If answers to both of these questions are available, then one can implement a retrieval system the properties and behavior of which are directly related to those of the two basic elements.

We begin with a body of knowledge $\mathcal{D}$ that consists of D data items $\{d_i \;/\; i = 1, \ldots, D\}$. Each data item is represented by a set of k properties and their respective values. These values can be categorical and/or numerical. Since categorical values can be remapped to numerical ones, we will concentrate, without loss of generality, on numerical data only. Notice that the remapping of categorical values to numerical values is possible but not necessarily a straightforward task. Much work has appeared in the literature on the subject, and a detailed discussion of it exceeds the scope of our presentation.

Each data item and its associated k-tuple property values can be thought of as a point in a k-dimensional space of representations. Data items that are "similar" have corresponding points that are "close" to each other in the representation space. Appropriate expressions have been devised that can evaluate data-item similarity and quantify it through the computation of a "distance" between the items' point-representations. A time-honored observation is that there is no such thing as a universal representation scheme: over the years, domain-specific representations have been proposed that were more or less successful in tackling domain-specific tasks. Typically, representations varied as a function of the tasks that were considered.

With the k-dimensional representations for each of the D data items fixed, the next issue is that of storing them in an appropriate data structure that supports operations such as insertion of a new item, deletion of an existing item, and matching of a given query to one or more items contained in the database. By matching here we mean determining those data items in the database that share properties (i.e., are similar) with the query. Several suggestions for the data structure have been made in the past, including linear arrays, R*-trees, hash tables, and so forth. Depending on the nature of the employed data structure, different characteristics are obtained in terms of space and time requirements.

Frequently, the data items contained in the database are allowed to undergo certain types of transformations. It is necessary that the matching operation be unaffected by the application of such transformations. This in turn requires that the employed k-dimensional representations remain *invariant* with respect to the possible transformations. Consider the simple example where the data items are Chinese tangram shapes. A representation scheme for each tangram that uses the raw coordinates of the corner points will not be a useful choice: simple translation of the shape will result in a completely new representation for the shape, that is, a new point in the k-dimensional space. Clearly, other schemes are required. When the representations remain invariant under the allowed transformations, then the matching step mentioned above will allow for a much larger class of queries to retrieve data items from $\mathcal{D}$.

To recapitulate, two elements need to be addressed: a representation scheme that will appropriately encode the various data items that comprise the body of available knowledge, and a storage/retrieval mechanism that operates on the representations and at the very least allows one to carry out the above operations. Here we assume that the distance metrics that are necessary for data item comparisons are an integral part of the representation element.

As far as storage/retrieval mechanisms are concerned, hash tables have been increasingly used in the last several years and have resulted in real-world working systems in many application domains. The two main characteristics of hashing schemes are large space requirements and exceptional performance: hash tables effectively trade space for computation.

As far as representation schemes go, they are, as we mentioned, a domain-specific consideration. Each particular instance depends on the exact nature of the data items and the properties that are being considered. And this is precisely at the center of our discussion in this chapter.

The context in which we have developed our framework is the following: we are given a compound/molecule C in the form of a set of coordinates of the compound's atomic sites. We are also given a database $\mathcal{D}$ of molecules that are described in terms of sets of coordinates of the atomic sites for each of the member molecules. Bonds connect the various atomic sites in both the query molecule and the molecule-members of the database. Some of the bonds are rotatable and thus allow for torsional flexibility. Torsional flexibility permits the groups of atoms rigidly attached at the two endpoints of a (rotatable) bond to rotate with respect to one another. A compound/molecule can contain one or more rotatable bonds, in which case it can assume any of an infinite number of conformations (3D configurations) via rotations around these bonds. Frequently, steric constraints or energy considerations will limit the number of choices, but the cardinality of the set of possible configurations will remain infinite nonetheless. This conformational flexibility of molecular structures opens a broad range of possibilities in the quest for potential ligands while at the same time rendering the problem much more difficult. In addition to the (internal) torsional flexibility, the molecules are allowed to undergo rigid transformations in 3D space; that is, each molecule as a whole can rotate and translate. In what follows, the compound/molecule C will be referred to interchangeably as "test compound," "test molecule," "query compound," "query molecule," or simply "query."

Given a query molecule C and a database $\mathcal{D}$ containing information about the 3D structure of a possibly large set of molecules, the following operations need to be carried out (Willett, 1987a; Martin et al., 1990): structure insertion, structure membership, substructure search, similarity search, and superstructure search. Note that structure membership can be implemented in terms of the substructure search operation. Furthermore, all of the search operations can be reduced to what we call "substructure similarity."

Substructure similarity refers to the result of an operation that, when given a compound C, a database $\mathcal{D}$, and a similarity measure d(.,.), allows the determination of all the compound-members of $\mathcal{D}$ that contain substructures that are similar to a substructure of C. The extent of the similarity between the molecules in question can be determined by the function d(.,.). The understanding here is that the implied common substructure may not necessarily be a proper subset of C. The similarity function d(.,.) will remain unspecified, but we assume it to be of a very general nature. Recall that the problem of substructure matching can be shown to be NP-complete (Garey and Johnson, 1979) by noting that it includes the problem of subgraph isomorphism as a special case. The computational complexity of the problem is further compounded by allowing torsional flexibility around the molecule's rotatable bonds. Finally, we should stress that this discussion concentrates on the problem of maximal atomic overlap between the query and molecule-members of $\mathcal{D}$: extensions of the described methodology exist that address the issue of similarity in terms of other properties (e.g., volumetric, electrostatic, etc.) and are not described here.

The inherent computational complexity of the substructure similarity task has typically plagued previously suggested methods. Even when one restricts the problem to the case of rigid molecules with no rotatable bonds, the problem remains computationally very demanding because of its 3D nature. The various techniques that have been proposed over the years typically differ in the definitions of the representation [as well as associated similarity measure d(.,.)] and the storage schemes. It is fair to say that most approaches have been computationally very demanding and did not scale well with database size.

For example, in the "atom-mapping" method (Pepperrell et al., 1990) the Tanimoto coefficient is computed using the result of pairwise comparisons of the rows from the distance matrices of two molecules, and used as an entry to an intermolecular similarity matrix. A greedy algorithm subsequently establishes the degree of similarity between any two molecules. The calculation is carried out for all pairs $\{C, M\}$ where C is the query and M is a molecule in the database.

In the "clique-detection" method (Kuntz, 1992), a number of different orientations are generated for each molecule M in the database prior to comparing it to the query C. Each of M's orientations is then overlaid on C and scored based on the presence or not of M's atoms in the vicinity of an atom of C; poorly scoring orientations are discarded. Pairs $\{C, M\}$ are examined iteratively, and the last n best scores are retained. This technique is at the core of the Mosaic molecular modeling system (Moon and Howe, 1990).

Molecular structures have been alternatively represented as connection tables and viewed as graphs (Willett, 1987b). The vertices of each such graph corresponded to the molecule's atomic sites. Two vertices were connected with a bond if and only if the corresponding atomic sites were connected with a bond. This representation allowed one to reduce the problem of searching for similar substructures to the search for subgraph isomorphisms, a problem that, as we have mentioned already, is NP-complete. Similarity measures for comparing molecular substructures were also developed based on graph-theoretic results (Dubois et al., 1991). More recent work presented evidence for the usefulness of a backtracking search algorithm enhanced with the "refinement procedure" heuristic (Ullmann, 1976). In a variation of the above scheme, the molecules of the database $\mathcal{D}$ are clustered as follows: using a similarity measure d(.,.), intermolecular similarities were calculated and the various molecules were assigned to clusters based on the scores of the pairwise comparisons. A query molecule C would be classified by identifying the cluster in which it belongs. The molecules that best match the query were drawn from the identified cluster as well as other neighboring clusters (Willett, 1988).

But one should not lose sight of the fact that treating the database molecules as rigid facilitates the matching operation at the expense of discarding large numbers of valid candidates: although the stored conformation of a molecule may not exhibit the pharmacophoric pattern/model sought, a different conformation of the same molecule may be biologically active. The conformational flexibility of molecular structures opens a broad range of possibilities in the quest for potential matches/answers while at the same time imposing a serious computational burden.

A straightforward extension of the initially suggested methods to accommodate flexible matching was the augmentation of the databases through the inclusion of a large number of representative conformations instead of a single one. This typically resulted in very large databases and longer searching times. A compromise called for storing only a handful of the possible conformations.

Some approaches suggested putting the flexibility into the database, for example, the Concord 3D system (Tripos Associates, St. Louis, Mo.). An alternative scheme put the conformational flexibility into the search. Examples of this latter approach are the Chem-X system (Chem-X Chemical Design, Mahwah, N.J.) and the system discussed in Murrall and Davies (1990). A third suggestion put the flexibility in the query itself (Christie et al., 1990; Güner et al., 1990). The query in this case combined both rigid and flexible components and was iteratively refined by searching a database of compounds with "known" activity until the desired selectivity is obtained. Once the final query was available, it was used to search a database of compounds with "unknown" activities to identify potential leads. A comparison study (Haraki et al., 1990) showed that augmenting a database with multiple conformations of a given molecule generally enhances a method's performance. But at the same time, the resulting effectiveness largely depends on the method used to generate the various conformations to be added in the database.

As an alternative to the multiple inclusion of a molecule in the database, a certain type of minimization could be carried out in "discrepancy space" (Hurst, 1994). This approach was much faster but also inherited all of the problems of nonlinear optimization approaches.

The more successful search techniques attacked the problem of conformational flexibility in a computationally demanding way. Any attempts to reduce the burden through the incorporation of heuristics had a direct impact on the quality of the produced results: valid matches could now be missed. It should be pointed out, however, that some of the proposed heuristics proved to be of general applicability and could be used in settings where conformational flexibility was not one of the parameters of the problem.

The use of descriptors as database screens has been at the core of many proposed techniques. For example, the work in Nilakantan et al. (1993) introduces a two-stage method that in essence characterizes shape without the need for examining a multitude of docking orientations. A number (signature) is used to capture geometric characteristics that are particular to each molecule; however, due to the way it is generated, the representation is not unique. During the second stage, a similar number is generated for the query molecule and compared with each of the stored signatures; molecule pairs whose scores exceed a certain threshold are then compared for intersection. Various descriptors (i.e., signatures) that cover a large range of molecular properties are presented and thoroughly studied in Fisanick et al. (1992, 1994).

It is fair to say that the above techniques did not scale well with the size of the database and/or did not fully exploit the constraints imposed by the descriptors to limit the extent of the search and the resulting computational burden.

On the other hand, hash-table-based approaches have been based on the identification of certain invariant descriptors carrying information about geometric constraints ("geometric hashing;" Lamdan et al., 1990; Rigoutsos, 1992). The descriptors can be used to store in a hash table a partial representation of a data item, in our case a molecule. Compatible molecules can be determined by computing the indices from a test input (Nussinov and Wolfson, 1991; Fischer et al., 1992a, 1992b; Sandak et al., 1995), retrieving the partial representation from the hash table, and integrating the evidence directly, thereby eliminating the need to scan the entire database for one or more matches.

Unlike the scan-based techniques, the class of hashing approaches has increased storage requirements. In particular, the various incarnations of the algorithm derive their speed by precomputing results and storing them in appropriately constructed hash tables. This precomputation can be performed off-line and is done only once; the results are stored on disk and used when needed. The hashing approach in essence trades space for computation; in the face of decreasing slow-storage costs, the trade-off is becoming increasingly justifiable and reasonable.

The method we propose belongs to the geometric hashing class of algorithms. It proceeds in two phases and also shares conceptual similarities with the one described in Wolfson (1991). During the first, off-line phase, and for each molecule in the knowledge base, a single conformer is identified; this conformer can be arbitrary (or can be identified in an educated manner), but any conformer from the infinite set of conformers for the molecule under consideration can be selected. For this conformer, multiple entries are made into a hash table, and they encode implicitly the molecule's ability to assume new conformations via rotations about its rotatable bonds. Note that for each molecule in the database, the only entries made are those generated by the single conformer that has been identified—this results in savings when compared to methods that generate multiple representations for each of a number of conformers. In the second phase, which takes place on-line, the stored implicit information is exploited in a manner reminiscent of the pose-clustering method (Stockman, 1987) and hypotheses are produced that correspond to placements of the molecule's rigid groups in the 3D space. We use the knowledge of the connectivity properties for the rigid groups of each molecule-answer to combine (thread) the hypotheses into full-fledged conformations (poses). No search of the molecule's conformational space is required during the threading, making this phase very efficient. Moreover, those database molecules that exhibit maximal atomic overlap with the query molecule are reported in the appropriate conformation.

In the following section, we describe the representation scheme that facilitates the speedy 3D substructure matching in the presence of rotatable bonds. In section 7.2, we describe the storage and retrieval phases of the algorithm, while in section 7.3 we summarize the algorithm and outline its properties. Finally, section 7.4 describes some preliminary results obtained on a public domain database. Conclusions are discussed in section 7.5.

7.1   The Representation Scheme

Before proceeding, recall the properties of our method. Given a query molecule C and a database $\mathcal{D}$ of molecules containing rotatable bonds, the method will determine those database members that share 3D substructures with the query to an extent exceeding a user-specified threshold; the method will also derive the conformer of each molecule-answer that will maximize the intersection of each suggested answer-pair (i.e., query + database molecule). And this will be achieved in the presence of rotatable bonds in either molecule of the answer-pair. An implicit assumption here is that the common substructures may not necessarily be proper subsets of the query molecule C.

Generally speaking, our work is a hybrid between pose clustering (Stockman, 1987) and geometric hashing. Indeed, the method uses a hash table to quickly generate hypotheses about those positions and orientations that would reveal any matching substructures between the test and database molecules. The generated hypotheses are then clustered together using a variation of the pose-clustering approach, but unlike the pose-clustering approach, our method does not require the exhaustive search of the information pertaining to all of the database molecules.

A novel element of this work lies in the representation scheme employed, allowing the generation of invariants that are unaffected by the presence of torsionally flexible bonds. A second novel element has to do with the fact that the proposed representation scheme treats uniformly both the rigid and the torsionally flexible molecule cases. Finally, the third novel element is the manner in which the hypotheses for the various rigid groups are combined to form a hypothesis for the full molecule.

Refer to the molecule of figure 7.1. The atomic sites of the molecule are labeled with letters A through S. Atoms of the molecule are connected with atomic bonds that are denoted by the labels of the atoms they connect; for example, the bond M–O exists between the atoms whose labels are M and O, respectively. The molecule consists of three rigid groups denoted rigid groups 1, 2, and 3, respectively, or rg1, rg2, and rg3 for short. The rigid groups are connected with one another via the rotatable bonds A and B (rbA and rbB, respectively) that allow for the torsional flexibility of the molecule.

The coordinates of the various atomic sites are expressed in a global coordinate frame. Atoms such as P are not considered to be separate from the group {G, H, I, J, K, L, M, N, O} despite the fact that the bond N–P may be rotatable. The reason is that any rotation around the bond N–P will not change the position of P in the global coordinate system. Moreover, any such rotation also will not change the relative placement of P with respect to the group of G through O.

It is easy to see that defining the coordinates (x, y, z) of three or more (noncollinear) atoms in the global coordinate frame suffices to define a global position and a global orientation for the rigid molecule in the global coordinate frame. Furthermore, the selected three sites define a (possibly skewed) local coordinate frame. Every remaining atomic site of the molecule has a set of coordinates (i.e., a position) in the local coordinate frame; these coordinates remain unchanged when the molecule undergoes a 3D rigid transformation (rotation and translation). Each of the two rotatable bonds allows the rigid group on one of its ends to rotate with respect to the neighboring group. Obviously, triplets comprising atoms that belong to more than one rigid group have structural properties that, in general, will change when one of these rigid groups rotates with respect to the other.

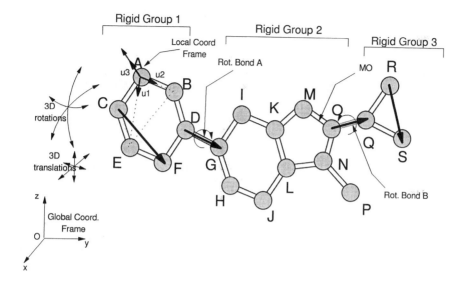

Figure 7.1   A general molecule with torsionally flexible bonds.

Since such triplets of atoms do not exhibit structural properties that are fixed, they are not useful in the formation of representational invariants.

Having stated this, let us concentrate for the moment on rigid group rg1 and ignore the rotatable bond rbA emanating from it. It should be clear that if we augment this rigid group by including atom G, the new group will still be rigid. Indeed, notice that independent of any rotation around the bond D–G, the position of atom G relative to any of the atoms {A, B, C, D, E, F} remains unchanged. We refer to this new set of atoms as the augmented rigid group 1 or augmented-rg1. We can now view augmented-rg1 as a rigid molecule in its own right. An analogous situation exists for rg2 and rg3: rg2 can be augmented to include atoms D and Q, whereas rg3 can be augmented to include atom O.

We define the *magic pair* of an augmented rigid group as a pair of vectors whose endpoints are atomic sites from this group. The two vectors should not be collinear and may share a tail (or a head).

- If there is a single rotatable bond emanating from an augmented rigid group, this bond will be one of the vectors in the magic pair: the second vector will comprise two atomic sites from the augmented group and cannot be parallel to the first.
- If there are exactly two rotatable bonds emanating from an augmented rigid group, then the magic pair will comprise these two bonds. Notice that the two rotatable bonds are assumed to be noncollinear. Whenever this noncollinearity assumption does not hold, then *two* magic pairs are associated with the augmented rigid group: each pair

contains one of the two bonds and the two magic pairs also share a vector that is not parallel to the (common) direction of the two rotatable bonds.

- If there are more than two rotatable bonds emanating from an augmented rigid group, then more than two magic pairs are associated with the group: each magic pair comprises two of the rotatable bonds that are noncollinear.

- Finally, if the molecule has no rotatable bonds, then a magic pair is associated with the only rigid group present: the two vectors can be formed by arbitrarily selecting atomic sites from the molecule and should not be collinear.

In any case, the vectors forming the magic pair are called *magic vectors.*

For example, in the molecule of figure 7.1, a possible magic pair for augmented-rg1 is C → F and D → G where X → Y represents a vector with the two endpoints being atom X and atom Y, respectively. For augmented-rg2, the magic pair comprises D → G and O → Q. Finally, a possible magic pair for augmented-rg3 is O → Q and R → S. All of these choices are shown in figure 7.1. In the general case, a magic vector for a (augmented) rigid group can be defined using endpoints that are uniquely determined as a *function* of the coordinates of the atoms comprising the (augmented) rigid group. Finally, note that given the two endpoints of a magic vector, we can impose a direction on the vector by using the lexicographic ordering of the respective labels. Of course, other choices are possible as well.

Why do we define the magic pairs with the help of rotatable bonds? First, it is important to realize that positioning the pair of magic vectors in the 3D space immediately positions the associated rigid group; this is a direct consequence of the way the vectors were constructed. Given that the magic vectors are noncollinear, they form a coordinate frame, the magic coordinate frame. For simplicity, the frame is assumed to have its origin at the center of mass of the four endpoints defining the respective magic vectors. The frame's x and y principal axes are defined with the help of the two magic vectors. The noncollinearity assumption guarantees that the z-axis of the frame can be formed. Finally, it should be pointed out that the magic coordinate frame will be skewed in the general case. The magic pair is by definition rigidly attached to the associated rigid group and thus moves together with the group whenever the latter undergoes rigid transformations. Consequently, each of the magic vectors has a fixed position and orientation in any *local* coordinate frame that can be defined by means of three (noncollinear) atoms belonging to the group. That is, for each magic vector, there are translation vectors and rotation matrices that uniquely determine the position of its two endpoints in a local coordinate frame. Conversely, given the definition of any one local coordinate frame and the appropriate set of translation vectors and rotation matrices, the magic pair associated with the group can be reconstructed unambiguously. Moreover, since the same three sites that define the local coordinate frame have a fixed position and orientation in the global coordinate frame, the translation vectors and rotation matrices allow the unique determination of the position and orientation of the magic pair (and thus of the respective rigid group) in the global coordinate frame.

Second, note that any two rigid groups that share a rotatable bond will also share a magic vector; this is the vector that corresponds to their common bond. During the recognition phase, we will be joining together rigid groups that are supported by the generated hypotheses in order to form larger groups: having the magic pairs of two consecutive groups share a magic vector will allow us to "hop" from a given rigid group to the next if and only if both rigid groups compute the same 3D position and orientation for the bond that connects them. All of this will become evident as we describe the threading stage of the method.

Having discussed the representation scheme, we next describe the storage scheme.

7.2 Storage/Retrieval and Threading

With the representation scheme described above, a given molecule with N atoms has its rigid groups identified and augmented. For each augmented rigid group, a magic pair involving at least one of the rotatable bonds emanating from the group is identified and fixed. Next, triplets of noncollinear atomic sites can be formed; at most, $\binom{N}{3}$ unordered and $3! \times \binom{N}{3}$ ordered triplets can be formed. To each formed triplet we associate a set of translation and rotation matrices representing the position and orientation of the rigid group's associated magic pair in the respective local coordinate frame. (Alternatively, any other equivalent representation of the magic pair can be used.)

The association of each local coordinate frame with the appropriate translation and rotation matrices is accomplished with the help of a hash table. The hash table consists of a number of hash bins that can be indexed by means of the properties (geometric and/or others) of the triplet of atoms forming the local frame. Given a triplet of atoms, an index is generated and used to identify and access a location (i.e., hash bin) of the hash table during storage as well as retrieval. An entry is made at the accessed hash bin that contains information about the identity of the molecule, the identity of the rigid group under consideration, and the translation and rotation matrices (or equivalent information) that will allow the reconstruction of the magic pair; other pertinent information may also be included. A given hash bin will in general contain more than one such entry, each of which corresponds to one or more distinct molecules.

The procedure described above is repeated for all the rigid groups of the molecule under consideration and for all the molecules in the database $\mathcal{D}$.

For the purposes of the method we have described, a minimum requirement for the set of characteristics would be the inclusion of the geometric properties of the triplet. Refer again to figure 7.1 and consider the triplet of atoms A–E–B. For simplicity, we assume that the triplet is ordered. This triplet forms a triangle, the three sides of which have lengths $\ell_i, i = 1, 2, 3$, and the angles at the three vertices are equal to $\theta_i, i = 1, 2, 3$, respectively. By appropriately quantizing the values of any subset of these properties, we can produce an index that is particular to the triplet under consideration. One subset that we found worked well when we experimented with the public part of the National Cancer Institute's database of molecules that were tested for carcinogenic activity was ℓ_1, ℓ_2, and θ_1.

It is clear that additional information can also be used to form the index. This information may be capturing other properties (e.g., physical, chemical, etc.) pertaining to the atoms forming the tuple and/or the molecule to which the atoms belong. For example, a value can be attached to each of the three atoms A, E, and B that corresponds to their chemical type; this value can be used as an additional element of information when forming the index of a hash bin. Incorporating more information in the index creation makes the index more descriptive. However, and depending on the nature of the problem at hand, this increase in discrimination power may occur at the cost of the resulting sensitivity; thus, caution should be exercised when trying to strike a balance between these two requirements. An issue related to the index generation scheme is that of the resulting hash bin occupancy. Typically, components of an index such as ℓ_1, ℓ_2, and θ_1 are correlated. Consequently, there is a preference for certain regions of the associated hash table. In other words, the probability of making an entry in a given hash bin depends on the bin and is not constant for all bins. This in turn translates into hash tables with bins containing varying numbers of entries. The observation is not particular to the way of producing indices described above, but instead reflects a property endemic to all indexing schemes. In fact, it has been observed for a variety of data types and in several problem domains. There are a number of ways in which this nonuniform distribution can be alleviated, but they exceed the scope of this discussion (see Rigoutsos and Delis, 1996, for a detailed analysis). For an alternative indexing scheme that uses only geometric information and guarantees a constant expected occupancy across the entire set of hash bins, refer to Rigoutsos (1997).

Recall that the population of the hash table occurs in the preprocessing phase; as such it is carried out only once and takes place off-line. On the other hand, the retrieval phase takes place on-line, and this is where efficiency is essential. The use of hash tables guarantees that all relevant hypotheses will be retrieved and formed in a speedy manner.

When presented with a query molecule (possibly containing rotatable bonds), triplets of atoms are formed by selecting from among the atoms forming the query molecule. During the retrieval phase it is not necessary to distinguish rigid groups in the query molecule. Each of the formed atom triplets (and the atoms' properties, if applicable) will give rise to an index that will be used to access a bin in the hash table; for each of the entries in the accessed hash bin a hypothesis (Molecule-Id, Rigid-Group-Id, Group's-Magic-Pair-Position-and-Orientation) will be reconstructed.

After all of the query molecule's triplets have been exhausted, there will be a collection of hypothesized (Molecule-Id, Rigid-Group-Id, Group's-Magic-Pair-Position-and-Orientation) tuples. Note that not all of these tuples will be distinct. Indeed, a given tuple is likely to appear in the collection more than once. The degree of repetition is proportional to the number of atom triplets made up of atoms participating in the substructure that is shared by the query and the indicated rigid group of the molecule specified by the tuple: all these triplets will be reconstructing the same (within tolerance) hypothesis for the position and orientation of the magic pair for the candidate rigid group and molecule combination.

Having produced this collection of hypotheses for the positions and orientations of various rigid groups in different molecules, we next set out to answer the question of whether there is a substructure common to both the query and some torsionally flexible database molecule. It is a maximum-size substructure that we are seeking, and this substructure will generally span more than one rigid group.

It is clear that since the database molecules are assumed to be torsionally flexible, a maximum-size common substructure may reveal itself only if the database molecule is placed in the appropriate conformation. Each element of the collection of the hypothesized tuples (Molecule-Id, Rigid-Group-Id, Group's-Magic-Pair-Position-and-Orientation) in essence states the following: if the rigid group Rigid-Group-Id of molecule Molecule-Id is positioned in the 3D space so that its associated magic pair assumes the position and orientation indicated by the tuple's Group's-Magic-Pair-Position-and-Orientation, then the rigid group Rigid-Group-Id of the molecule Molecule-Id will have maximal overlap with a part of the query molecule; the amount of structural overlap will be analogous to the number of atom triplets that have voted for the tuple in question.

But each of these tuples provides additional information—indeed, they each dictate the position and orientation of the respective rigid group's magic pair. Refer back to rigid group rg2 from figure 7.1. Recall that we have defined the magic pair for this group by means of the two rotatable bonds emanating from it. Any hypothesis tuples produced for this group will in essence dictate possible position and orientation choices for rotatable bonds rbA and rbB and will reveal any existing structural overlap between rg2 of this molecule and the query. Analogous statements can be made for the hypothesis tuples generated for rigid groups rg1 and rg3. Any structural match between the query and rigid group rg2 can be extended to include also any existing structural match between the query and rigid group rg3 if and only if there is a hypothesis (Molecule-Id, rg2, Group's-Magic-Pair-Position-and-Orientation) as well as a hypothesis (Molecule-Id, rg3, Group's-Magic-Pair-Position-and-Orientation) both placing the common rotatable bond rbB (= shared magic vector) in the same 3D position and orientation. Similarly, hypotheses for rg2 can be combined with compatible hypotheses for rg1 to extend the structural intersection between this molecule and the query.

In figure 7.2, we show this situation schematically by showing hypotheses (Molecule-Id, Rigid-Group-Id, Group's-Magic-Pair-Position-and-Orientation) for a three-rigid-group molecule such as that of figure 7.1. We depict each rigid group as an irregular closed contour. Attached to each group's contour are the hypothesized positions and orientations of the vectors of the magic pair associated with the rigid group.

Hypotheses referring to two distinct rigid groups connected through a rotatable bond are consistent with one another if and only if they both agree on the 3D position and orientation of the magic vector corresponding to the common rotatable bond. If a common substructure exists between the query molecule and the molecule under consideration that spans all the rigid groups of the latter, then a path such as the one shown in figure 7.2 must exist. Such a path has the following properties:

- each hypothesis participating in the path corresponds to a position and an orientation of the respective rigid group that is consistent with the position and orientation of any other neighboring group;
- hypotheses corresponding to two rigid groups that are connected by a rotatable bond place the respective groups in the 3D space in a way that agrees with the presence of the (shared) rotatable bond;
- the path itself corresponds to a valid conformation of a database molecule that exhibits the maximum possible structural overlap with the query molecule; and
- if for no possible conformation of the database molecule does a common substructure exist that spans all of the molecule's rigid groups, then no path will be complete: a partial path is a path that comprises hypotheses that do not include *all* of the rigid groups of the molecule.

Figure 7.2 also shows how the magic pair is selected and fixed for each group. In particular, and as mentioned above, the hypotheses for rigid groups rg1 and rg3 involve magic vectors that are internal to the groups: unlike rigid group rg2 of the molecule in figure 7.1, which has two (non-collinear) rotatable bonds emanating from it, rigid groups rg1 and rg3 have only one bond. For such groups, the second vector of the magic pair will be defined using atoms other than those participating in the rotatable bond; in fact, any vector that is not collinear with the one rotatable bond emanating from the unit will suffice.

7.3   The Algorithm

In the preceding sections, we outline the representation scheme that we are proposing and also show how it can be used in the storage and retrieval phase of the system to carry out a 3D-substructure matching search of a

Hypotheses
For Rigid Group 1

Hypotheses
For Rigid Group 2

Hypotheses
For Rigid Group 3

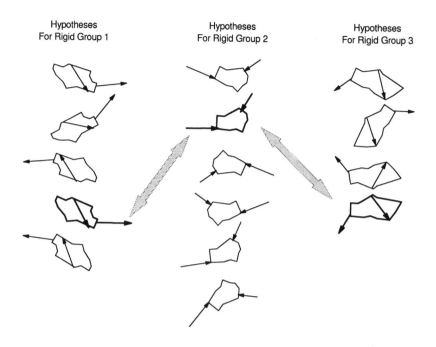

Figure 7.2   Combining hypotheses from multiple rigid groups.

given query molecule in a database $\mathcal{D}$ of known molecules. Here we recapitulate the various steps.

The method has two phases, similar to the traditional geometric hashing that is described in Nussinov and Wolfson (1991). The first phase (= storage) is a preprocessing phase; it takes place off-line and operates on the database of known molecules, and its end result is the population of the employed hash table with redundant representations of partial substructures from the known molecules. The molecules in $\mathcal{D}$ are assumed to be torsionally flexible.

The storage phase begins with the selection of a molecule M from $\mathcal{D}$. All the rigid groups of M are identified next, together with the rotatable bonds emanating from each one of them. Each of the rigid groups of M is processed in turn: for the rigid group under consideration, we first identify one magic pair (or more, depending on the molecule's topology) and associate it with the rigid group; we then generate the *reference-tuple selection set* comprising all atomic sites that will participate in tuples generating the various indices. (These tuples are typically triplets, but higher order tuples are also possible and have been used in our work. To simplify the description we assume that only triplets are formed.)

At this step, we also carry out group augmentation with the incorporation of atoms belonging to other groups and participating in rotatable bonds that emanate from the group under consideration. By subselecting from this set, we produce a number of noncollinear triplets. We use each such triplet as the core for a reference frame in which the magic pair's position and orientation are expressed; this representation is in terms of translation and rotation matrices. Using the properties (geometric and/or other) of the formed triplet, we generate the index of the respective hash bin, and the magic pair's representation is incorporated in an entry to that bin. The other components of the entry include the identities of the molecule and rigid group under consideration, identifiers for the atoms forming the triplet, and possibly other information.

We repeat the above process as necessary until all of the triplets, rigid groups, and database molecules have been considered. On termination, the hash table contains all of the rigid groups and molecules *classified* according to the attributes of the tuple used to determine the hashing index.

The retrieval phase of the algorithm takes place on-line and assumes the availability of the hash table that was generated during the preprocessing. A query molecule is presented to the system and initiates the substructure matching process.

Although it is not necessary, determining any rotatable bonds that are present in the query molecule as well as the resulting rigid groups may generate computational savings. Whether or not this step takes place, one or more collections of atoms rigidly connected with one another will be identified in the query molecule. The retrieval proceeds with the selection of one of the identified rigid groups and the formation of the *matching-tuple selection set* that includes all of the atoms that can participate in the creation of triplets (or higher order tuples if appropriate). A triplet is selected and the respective (skewed) local coordinate frame is formed. Using the properties of the selected triplet, an index into the hash table is produced in a manner similar to that used during storage. The accessed bin will, in general, contain multiple entries: for each of the entries in the hash bin, the representation of the magic pair(s) contained therein is used in conjunction with the active local reference frame to produce a hypothesis in a voting table. The hypothesis is about the position and orientation (in the *global* coordinate frame) of a database molecule's rigid group (see also figure 7.2). The above steps are repeated until all the triplets in all of the identified groups of the query molecule have been exhausted.

On termination, the voting table likely contains entries with a multiplicity higher than 1. These entries will correspond to multiple frame tuples that corroborate a certain placement of a given molecule's rigid group in the global coordinate frame. The multiplicity of occurrence reflects the extent of similarity (i.e., the size of the common substructure) between the query molecule and the named rigid group of the named molecule. A simple histogramming operation of the voting table's contents can determine the identities of the rigid groups and the respective molecules in order of decreasing support; attached to each such answer is the position and orientation of the group's magic pair that will place the group in the best overlapping registration with the query molecule.

At this point of the retrieval phase, we have answered the question of which database molecules' rigid groups have common substructures with the query molecule. It is possible that some molecules of the database have multiple rigid groups sharing substructures with the query molecule. In such cases, we can actually derive the conformation of the database molecule that is compatible with the individual results returned by the histograming. In particular, assume that a database molecule with identifier m has more than one rigid groups sharing substructures with the query molecule. We subselect the corresponding answers from the histogram and partition them into categories: the ith category will contain the answers corresponding to the ith rigid group of m (see also figure 7.2). These answers can now be treated as *hypotheses* for the purpose of producing a conformation for the molecule m that will exhibit the maximum possible overlap with the query molecule.

Recall that we have knowledge of m's topology and thus know which rigid groups are connected with one another. A hypothesis from the ith category can be augmented and combined with a hypothesis from the jth category if and only if (a) the ith and jth rigid group of molecule m are connected, and (b) both hypotheses agree on the position and orientation of the rotatable bond that they share. Joining these two hypotheses will "consume" the magic vector corresponding to the common rotatable bond and will generate a partial conformer involving rigid groups $\{i, j\}$; this partial conformer can be augmented further from, say, its i end through combination with a hypothesis from, say, the kth category if and only if (a) the ith and kth rigid group of molecule m are connected, and (b) the newly selected hypothesis agrees on the position and orientation of the rotatable bond connecting groups i and k. This will again consume the common magic vector and generate a partial conformer involving rigid groups $\{k, i, j\}$. This procedure continues until either a full conformer involving all of m's rigid groups is built or the partial conformer under consideration cannot be augmented further.

Note that the list of *chained* (or *threaded*) hypotheses suffices to build a conformer for m: the respective magic pairs need to be placed in 3D space in the position and orientation with which they participate in the chain of hypotheses. The above steps can be repeated beginning with another hypothesis for m and for all database molecules that have hypotheses generated by the histograming step.

7.4   Experimental Results

In this section, we present some preliminary results from a prototype implementation of the representation scheme that we have described. The database we used was the publicly available segment of the National Can-

Table 7.1   The number of rotatable bonds present in database molecules.

Rotatable bonds	Number of molecules
0	11,865
1	13,341
2	17,791
3	17,569
4	16,557
5	11,637
6	9,502
7	5,994
8	4,906
9	3,154
10	2,500
≥ 11	8,249

cer Institute's database of substances that have been examined for carcinogenic activity (Milne et al., 1994). After excluding those entries that corresponded to disconnected molecules or had no 3D information listed, we generated a database containing 123,065 molecules. Table 7.1 shows the statistics in terms of the number of rotatable bonds present in database molecules. Note that a large number of molecules contained several rotatable bonds.

The hash table that was generated for this database using the above-described method was approximately 2.0 gigabytes. For this particular implementation, triplets of atoms were used to form tuples. Each triplet generated a *quadruplet* of numbers: ℓ_1, ℓ_2, θ, and c; c corresponded to one of 41 atom types (as outlined in Tripos's SYBYL system; see SYBYL, 1997) for one of the triplet's atoms. The produced quadruplet of numbers was subsequently quantized and used to generate the index of a hash bin where an entry was made. During the selection of triplets, we only allowed triplets whose respective triangles' sides had lengths between 1 and 4 Å. We made no attempts to optimize the hash table generation phase, which required approximately 10 hours on an 8-processor supercomputer.

For the recognition phase, we used molecules of varying sizes. The search times depended on the nature of the query and typically required only a few minutes. Again, an 8-processor supercomputer was used, with the hash table shared equally among the processors.

Finally, we should point out that several hot spots were identified in the hash table. Some were the result of correlation between the indices, but most were primarily the result of overrepresented structures that were present in the database (e.g., rings, ring systems, etc.). This observation suggests the need for alternative multilevel descriptions and/or representation schemes. Both of these areas are currently being researched.

7.5   Discussion

We have introduced and discussed a new method for representing and describing 3D molecular structures in a manner that facilitates 3D substructure matching searches in the presence of torsional flexibility in the molecules of the database.

The proposed representation scheme is part of a two-phase method that uses a hash table. During storage, the hash table is used to store redundant representations of small molecular structures (triplets of atoms). Despite the fact that the molecules are allowed to be torsionally flexible, and thus can assume an infinity of possible conformations, only one conformer per database molecule is needed to generate hash table entries.

During the retrieval/matching phase, the hash table is used to propose and cluster rigid-group-placement hypotheses in a manner reminiscent of the pose-clustering technique. The use of the hash table guarantees the fast generation of these hypotheses as well as substantial filtering of the possible candidate answers. Consistent hypotheses can be joined together to form larger substructures shared by the query and database molecules (augmentation step).

What is most important is that this augmentation step can take place at a substantial fraction of the time needed to sample a molecule's conformation space. We wish to stress that no conformations are ever sampled for a given molecule-answer. Instead, the task at hand is reduced to a set of consistency checks for already available, valid hypotheses, and it consequently is very efficient computationally. The end result of this augmentation step is a conformation for the molecule-answer with no torsional-resolution limitations; methods that rely on either explicit or implicit sampling of the conformation space using intervals that follow a deterministic schedule in essence impose a quantization step. The resulting discrimination power of any such method is quantized in that it collapses more than one valid conformation into one representative. To contrast this, our method's resolution ability is limited only by machine precision.

We have found that the method described here is suitable for databases of small molecules (e.g., drug molecules) where the exhibited flexibility can be captured by the torsionally flexible bond model. Through our building of a prototype system, we have also identified areas where further research is warranted. One such area is related to the observed nonuniform occupancy of the various hash bins and is currently being pursued because of its impact on the scalability of the system to databases containing a million or more molecules.

Acknowledgments

We thank the National Cancer Institute in Bethesda, Maryland, for kindly providing the public segment of their database of substances that have been tested for carcinogenic activity.

Part III. System Components for Discovery

Chapter 8

A Framework for Biological Pattern Discovery on Networks of Workstations

Bin Li, Dennis Shasha, and Jason T. L. Wang

Biological pattern discovery problems are computationally expensive. A possible technique for reducing the time to perform pattern discovery is parallelization. Since each task in a biological pattern discovery application is usually time-consuming by itself, we might be able to use networks of workstations (NOWs) that communicate infrequently.

Persistent Linda (PLinda) is a distributed parallel computing system that runs on NOWs and it automatically utilizes idle workstations (Anderson and Shasha, 1992; Jeong, 1996). This means that labs can do parallel pattern discovery without buying new hardware.

We propose an acyclic directed graph structure, exploration dag (E-dag for short), to characterize computational models of biological pattern discovery applications. An E-dag can first be constructively formed from specifications of a pattern discovery problem; then an E-dag traversal is performed on the fly to solve the problem. When done in parallel, the process of E-dag construction and traversal efficiently solves pattern discovery problems. Parallel E-dag construction and traversal can be easily programmed in PLinda.

8.1 Biological Pattern Discovery

Finding active motifs in sets of protein sequences and in multiple RNA secondary structures are two examples of biological pattern discovery. Before discussing the framework, we introduce these two applications and briefly describe their computational models.

8.1.1   Discovery of Motifs in Protein Sequences

Consider a database of imaginary protein sequences $\mathcal{D} = \{$FFRR, MRRM, MTRM, DPKY, AVLG$\}$ and the query "Find the patterns P of the form $*X*$ where P occurs in at least two sequences in $\mathcal{D}$ and the size of P $|P| \geq 2$." (X can be a segment of a sequence of any length, and $*$ represents a variable length don't care [VLDC].) The good patterns are $*$RR$*$ (which occurs in FFRR and MRRM) and $*$RM$*$ (which occurs in MRRM and MTRM).

Pattern discovery in sets of sequences concerns finding commonly occurring subsequences (sometimes called *motifs*). The structures of the motifs we wish to discover are regular expressions of the form $*S_1 * S_2 * \ldots$ where $S_1, S_2, \ldots$ are *segments* of a sequence, that is, subsequences made up of consecutive letters, and $*$ represents a VLDC. In matching the expression $*S_1 * S_2 * \ldots$ with a sequence S, the VLDCs may substitute for zero or more letters in S. Segments may allow a specified number of mutations; a mutation is an insertion, a deletion, or a mismatch.

We use terminology proposed in Wang et al. (1994a). Let $\mathcal{S}$ be a set of sequences. The occurrence number of a motif is the number of sequences in $\mathcal{S}$ that match the motif within the allowed number of mutations. We say the occurrence number of a motif P with respect to mutation i and set $\mathcal{S}$, denoted $occurrence\_no_{\mathcal{S}}^{i}(P)$, is k if $*P*$ matches k sequences in $\mathcal{S}$ within at most i mutations, that is, if the k sequences contain P within i mutations. Given a set $\mathcal{S}$, we wish to find all the active motifs P where P is within the allowed Mut mutations of at least $Occur$ sequences in $\mathcal{S}$ and $|P| \geq Length$, where $|P|$ represents the number of the non-VLDC letters in the motif P. (Mut, $Occur$, $Length$, and the form of P are user-specified parameters.)

8.1.2   Discovery of Motifs in RNA
Secondary Structures

Finding approximately common motifs (or *active motifs*) in multiple RNA secondary structures helps to predict secondary structures for a given mRNA (Le et al., 1989; Zuker, 1989) and to conduct phylogenetic study of the structure for a class of sequences (Shapiro and Zhang, 1990). Adopting the RNA secondary structure representation proposed in Shapiro and Zhang (1990), we represent both helical stems and loops as nodes in a tree. Figure 8.1 illustrates an RNA secondary structure and its tree representation. The structure is decomposed into five terms: stem, hairpin, bulge, internal loop, and multibranch loop. In the tree, H represents hairpin nodes, I represents internal loops, B represents bulge loops, M represents multibranch loops, R represents helical stem regions (shown as connecting arcs), and N is a special node used to make sure the tree is connected. The tree is considered to be an ordered one where the ordering is imposed based on the 5' to 3' nature of the molecule. This representation allows one to encode detailed

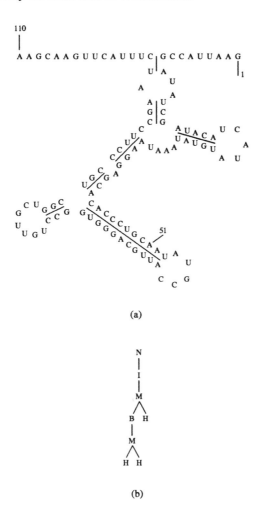

(a)

(b)

Figure 8.1 Illustration of a typical RNA secondary structure and its tree representation. (a) Normal polygonal representation of the structure. (b) Tree representation of the structure.

information of RNA by associating each node with a property list. Common properties may include sizes of loop components, sequence information, and energy.

We consider a motif in a tree T to be a connected subgraph of T, namely, a subtree U of T with certain nodes being cut at no cost. (Cutting at a node n in U means removing n and all its descendants, i.e., removing the subtree rooted at n.) The dissimilarity measure used in comparing two trees is the *edit distance*, that is, the minimum weighted number of insertions, deletions, and substitutions (also known as relabelings) of nodes used to

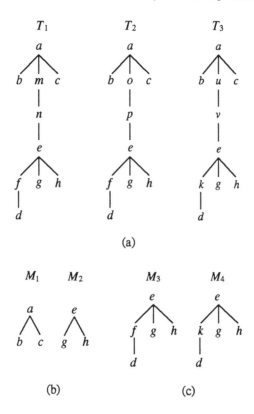

Figure 8.2   (a) The set S of three trees (these trees are hypothetical ones used solely for illustration purposes). (b) Two motifs exactly occurring in all three trees. (c) Two motifs approximately occurring, within distance 1, in all three trees.

transform one tree to the other (Shapiro and Zhang, 1990; Wang et al., 1994c). Deleting a node n makes the children of n the children of the current parent of n. Inserting n below a node p makes some consecutive subsequence of the children of p become the children of n. For the purpose of this work, we assume that all the edit operations have a unit cost.

Consider the set S of three trees in figure 8.2(a). Suppose only exactly coinciding connected subgraphs occurring in all the three trees and having size > 2 are considered active motifs. Then S contains two active motifs shown in figure 8.2(b). If connected subgraphs having size > 4 and occurring in all the three trees within distance 1 are considered active motifs, that is, if one substitution, insertion, or deletion of a node is allowed in matching a motif with a tree, then S contains two active motifs shown in figure 8.2(c).

We say a tree T contains a motif M within distance d (or M *approximately occurs* in T within distance d) if there exists a subtree U of T such that the *minimum* distance between M and U is $\leq d$, allowing zero or more

Table 8.1 A comparison of specifications of two biological pattern discovery applications.

	Pattern discovery	
	In protein sequences	In RNA secondary structures
Database	A set of sequences	A set of trees (representing RNA secondary structures)
Pattern	Partial sequence	Subtree
Good pattern	$Occurrence_{pattern}$ $> min\_occurrence$	$Occurrence_{pattern}$ $> min\_occurrence$
Task	Counting occurrences of *partial sequences* in subset of database	Counting occurrences of *subtrees* in subset of database

cuttings at nodes from U. Let S be a set of trees. The occurrence number of a motif M is the number of trees in S that contain M within the allowed distance. Formally, the occurrence number of a motif M with respect to distance d and set S, denoted $occurrence\_no_S^d(M)$, is k if there are k trees in S that contain M within distance d. Given a set S of trees, we wish to find all the motifs M where M is within the allowed distance $Dist$ of at least $Occur$ trees in S and $|M| \geq Size$, where $|M|$ represents the size, that is, the number of nodes, of the motif M. ($Dist$, $Occur$, and $Size$ are user-specified parameters.)

8.2   Biological Pattern Discovery Framework

In this section, we note the resemblance among the computational models of biological pattern discovery applications based on distance metrics. Table 8.1 shows a comparison of specifications of pattern discovery in protein sequences and in RNA secondary structures. Hereafter, we refer to pattern discovery in protein sequences simply as *sequence pattern discovery* and pattern discovery in RNA secondary structures as *tree pattern discovery*.

A *task* is the main computation applied on a pattern. Not only are all tasks of each application of the same kind, but also tasks of different applications are actually very similar. They all take a pattern and a subset of the database and count the number of objects in the subset that match the pattern.

The similarities among the specifications of these applications are obvious, which inspired us to study the similarities among their computation models. They usually follow a generate-and-test paradigm—generate a candidate pattern and then test whether it is active. Furthermore, there is some interdependence among the patterns that gives rise to pruning; that is, if a pattern occurs too rarely, then so will any superpattern.

8.2.1   Defining Biological Pattern Discovery Applications

In general, a biological pattern discovery application defines the following elements:

1. A database $\mathcal{D}$. In sequence pattern discovery, the database is a set of protein sequences; in tree pattern discovery, the database is a set of trees representing RNA secondary structures.
2. The function len(pattern p), which returns the length of p. The length of a pattern is a nonnegative integer. For example, in sequence pattern discovery, a pattern of length k is a sequence $*C_1 C_2 \ldots C_k *$ where $C_1, C_2, \ldots, C_k$ are letters. In tree pattern discovery, the length of a pattern is defined to be the number of nodes in the subtree.
3. A function $goodness$(pattern p), which returns a measure of p according to the specifications of the application. The goodness of a pattern p in sequence pattern discovery is the occurrence number of p in the set of sequences; the goodness of a pattern p in tree pattern discovery is the occurrence number of p in the set of RNA secondary structures.
4. A function $good(p)$, which returns 1 if p is a good pattern or a good subpattern and 0 otherwise. Zero-length patterns are always good. In both sequence and tree pattern discovery, a pattern p is good if $goodness(p) \geq$ some prespecified $min\_occurrence$.

Sometimes there are such additional requirements as minimum and/or maximum lengths of good patterns (e.g., in sequence pattern discovery). Without loss of generality, we disregard these requirements in our discussion unless otherwise noted.

Let us define an *immediate subpattern* of a pattern q to be a subpattern p of q where $len(p) = len(q) - 1$. Conversely, q is called an *immediate superpattern* of p. In sequence pattern discovery, immediate subpatterns of a length k pattern p are all $(k-1)$-prefixes and all $(k-1)$-suffixes of p. In tree pattern discovery, immediate subpatterns of a size k pattern p are all subtrees of size $k-1$, each of which can be derived by removing one node in p.

Thus, superpatterns of a pattern p are all of p's immediate superpatterns and their superpatterns. Similarly, subpatterns of a pattern p are all of p's immediate subpatterns and their subpatterns. Patterns of length 1 have the zero-length pattern as their only subpattern.

The result of a biological pattern discovery application is the set of all good patterns. If a pattern is not good, neither are any of its superpatterns. In other words, it is necessary to consider a pattern if and only if all of its subpatterns are good.

For all the patterns to be uniquely generated, a pattern q and one of its immediate subpatterns p have to establish a child-parent relationship (i.e., q is a *child pattern* of p and p is the *parent pattern* of q). Except for the zero-length pattern, each pattern must have one and only one parent pattern. For example, in sequence pattern discovery, *FRR* may be a child pattern of *FR*.

8.2.2 Solving Biological Pattern Discovery Applications

Having defined biological pattern discovery applications as above, it is easy to see that an optimal biological pattern discovery program does the following:

1. generates all child patterns of the zero-length pattern,
2. computes *goodness(p)* if all of p's immediate subpatterns are good,
3. if *good(p)* then generates all child patterns of p, and
4. applies steps 2 and 3 recursively until there are no more patterns to be considered.

Because the zero-length pattern is always good and the only immediate subpattern of its children is the zero-length pattern itself, the computation starts on all of its children, which are all length 1 patterns. After these patterns are computed, good patterns generate their child sets. Not all of these new patterns will be computed—only ones whose every immediate subpattern is good.

8.2.3 Exploration Dag

We propose to use a directed acyclic graph (dag) structure called *exploration dag* (E-dag for short) to characterize computational models of biological pattern discovery applications. We first describe how to map a biological pattern discovery application to an E-dag.

The E-dag constructed for a biological pattern discovery application has as many vertices as the number of all possible patterns (including the zero-length pattern). Each vertex is labeled with a pattern, and no two vertices are labeled with the same pattern. Hence, there is a one-to-one relation between the set of vertices of the E-dag and the set of all possible patterns. Therefore, we refer to a vertex and the pattern it is labeled with interchangeably.

There is an incident edge on a pattern p from each immediate subpattern of p. All patterns except the zero-length pattern have at least one incident edge on them. The zero-length pattern has an outgoing edge to each pattern of length 1. Figure 8.3 illustrates an E-dag mapped from a simple sequence pattern discovery application.

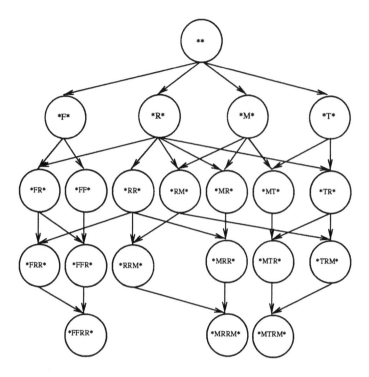

Figure 8.3 A complete E-dag for a sequence pattern discovery application on sequences FFRR, MRRM, and MTRM.

8.2.4   A Biological Pattern Discovery Virtual Machine

The E-dag structure enables us to define an efficient biological pattern discovery virtual machine. The input of the proposed virtual machine is a biological pattern discovery application with all the elements defined as above. The output of the virtual machine is the result of the biological pattern discovery application.

> **Definition 8.1** In an *E-dag traversal*, a vertex n is visited only if all vertices that have an incident edge on n have been visited.

There is plenty of pruning in an E-dag traversal. If a pattern p is not good, neither is any of its superpatterns. Therefore, there is no need to build the complete E-dag before we perform an E-dag traversal. Instead, an E-dag is lazily constructed—vertices are generated only when it is necessary to look at the patterns on them.

> **Theorem 8.1** For a biological pattern discovery application $\mathcal{A}$, an *E-dag $\mathcal{E}$ built from $\mathcal{A}$, and a sequential program $\mathcal{P}$ that solves $\mathcal{A}$ and*

that is optimal, an E-dag traversal of $\mathcal{E}$ is as efficient as an execution of $\mathcal{P}$ on input $\mathcal{A}$.

Proof sketch. It suffices to show that all patterns that the function *goodness*(p) would have been applied to are the same in an E-dag traversal of $\mathcal{E}$ as in an execution of $\mathcal{P}$.

According to the definition of E-dag traversal, patterns of length 1 are considered before any pattern of length 2. Because an execution of $\mathcal{P}$ would have to compute the goodness scores on all length 1 patterns as well and would obtain the same results as in an E-dag traversal, exactly the same length 2 patterns will be considered in both of them. This goes on until there are no more patterns to be considered. In other words, both will consider every pattern except those discarded by the subpattern-superpattern rule. This guarantees that an E-dag traversal is as efficient as an optimal program $\mathcal{P}$.

At the end of the computation, the resultant E-dag gives the results of the biological pattern discovery application. For both sequence pattern discovery and tree pattern discovery applications, for example, good patterns on the vertices of the resultant E-dag are active motifs.

Fact 8.1 *All E-dags built from the same input are isomorphic.*

This guarantees that the above discussion applies on all E-dags built for a single application.

8.2.5 A Parallel Biological Pattern Discovery Virtual Machine

We can now define a parallel biological pattern discovery virtual machine. The input of the parallel virtual machine is a biological pattern discovery application with all the defined elements. The output is the result of the biological pattern discovery application. The input and the output of the parallel virtual machine are exactly the same as those of the biological pattern discovery virtual machine presented earlier.

Definition 8.2 A *parallel E-dag traversal* is an E-dag traversal done in parallel.

Imagine that we have an unlimited number of workers each of which can individually work on any vertex of an E-dag. If, at the beginning of the E-dag traversal, there are n_1 patterns of length 1, n_1 workers can test these patterns in parallel. After all length 1 patterns have been tested, there are, say, n_2 patterns of length 2 to be tested; then n_2 workers can test these patterns in parallel; and so on.

8.3   Parallel Computing on Networks of Workstations

In practice, parallelizing biological pattern discovery computations is desirable, as databases to be mined are often very large and mining algorithms are usually computation intensive. We choose networks of workstations (NOWs) as the target platform, as opposed to massively parallel computers, due to their wide availability and cost-effectiveness. Unlike supercomputers installed in a few institutions, workstations are widely available; many institutions have hundreds of high-performance workstations that are unused most of the time. Second, they are already paid for and are connected via communication networks; thus, no additional cost is required for parallel processing. Third, they can rival supercomputers with their aggregate computing power and main memory.

The software we use to implement the parallel biological pattern discovery virtual machine is Persistent Linda (PLinda). PLinda is a set of upward extensions to Linda (Carriero and Gelernter, 1989), and is a distributed computing system designed to support fault tolerance in a heterogeneous environment and to exploit idle workstations. The design rationale behind PLinda is based on three observations:

- On NOWs, most machines are "private" and the owners of the workstations do not want to allow computation-intensive jobs to be run on their machines for fear of the degrading performance of their own processes. Therefore, it is crucial to guarantee that workstations will be used only while they are idle.
- It is necessary for a distributed computing system to run on networks of heterogeneous workstations because most institutions have heterogeneous workstations in their computing environments.
- When NOWs are used for parallel computing, the probability of failure grows as execution time or the number of processors increases. Since the suitable applications for networks of workstations are long running and coarse grained, fault tolerance is crucial for a distributed computing system that runs on NOWs. Without fault tolerance, a single component failure can cause an entire computation to be lost.

8.4   Parallel E-dag Traversal in PLinda

A parallel E-dag traversal PLinda program consists of a master and a worker. The master coordinates the whole computation and collects the results. A worker plays the role of a visiting worker described in section 8.2.5. Figure 8.4 (figure 8.5, respectively) is the pseudo-PLinda code of the master (worker, respectively).

```
child_pattern(zero-length pattern);
while (!done) do
   begin
      xstart;
      in(?node, ?score);
      if good(node) then child_pattern(node)
         else if (task_sent == task_done) then done = 1;
      xcommit;
   end;
out(poison tasks);
```

Figure 8.4 Pseudo-PLinda code of a parallel E-dag traversal master.

```
while (!done) do
   begin
      xstart;
      in("task", ?node);
      if !(poison task) then out(node, goodness(node))
         else done = 1;
      xcommit;
   end;
```

Figure 8.5 Pseudo-PLinda code of a parallel E-dag traversal worker.

The function *child_pattern(node)* generates all child patterns of the pattern on the input *node*, but it makes available to workers only those children whose immediate subpatterns are all good. As a result, a worker can take only patterns that are known to be necessary to be considered. Work tuples are of the form ("task," node). Goodness of a node is produced by a worker in a result tuple of the form (node, score). The function *good(node)* will determine if *node* is a good pattern according to goodness of all of its immediate subpatterns.

8.4.1   Experiments

We have applied our biological pattern discovery virtual machine and our parallel biological pattern discovery virtual machine on sequence pattern discovery. An E-dag traversal program in C programming language for sequence pattern discovery and from it a parallel E-tree traversal program in PLinda have been implemented. Figure 8.6 shows the experimental results of our sequence pattern discovery program running on 5, 10, 15, 20, 25, 30, 35, 40, and 45 Sun Sparc 5 workstations. Good speedup is achieved even when as many as 45 machines join the computation. For 15 and fewer machines, the speedup is particularly good.

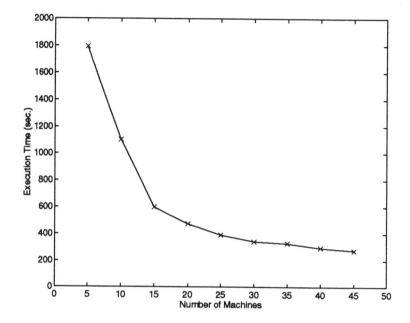

Figure 8.6   Running time of our parallel sequence pattern discovery program on 5, 10, 15, 20, 25, 30, 35, 40, and 45 machines. The protein database Cyclin has 47 sequences. Sequential running time is 4,686 seconds.

Further Information

The PLinda software for sequence pattern discovery is available from our Web site at http://merv.cs.nyu.edu:8001/~binli/pbpd/. Installation instructions and a PLinda user manual are also available at the site.

Chapter 9

Discovering Concepts in Structural Data

Diane J. Cook, Lawrence B. Holder,
and Gehad Galal

The large amount of data collected today is quickly overwhelming research-ers' abilities to interpret the data and discover interesting patterns. In re-sponse to this problem, a number of researchers have developed techniques for discovering concepts in databases. These techniques work well for data expressed in a nonstructural, attribute-value representation and address is-sues of data relevance, missing data, noise and uncertainty, and utilization of domain knowledge (Fisher, 1987; Cheeseman and Stutz, 1996). However, recent data acquisition projects are collecting structural data describing the relationships among the data objects. Correspondingly, there exists a need for techniques to analyze and discover concepts in structural databases (Fayyad et al., 1996b).

One method for discovering knowledge in structural data is the identifi-cation of common substructures. The goal is to find substructures capable of compressing the data and to identify conceptually interesting substruc-tures that enhance the interpretation of the data. Substructure discovery is the process of identifying concepts describing interesting and repetitive substructures within structural data. Once discovered, the substructure concept can be used to simplify the data by replacing instances of the sub-structure with a pointer to the newly discovered concept. The discovered substructure concepts allow abstraction over detailed structure in the orig-inal data and provide new, relevant attributes for interpreting the data. Iteration of the substructure discovery and replacement process constructs a hierarchical description of the structural data in terms of the discovered substructures. This hierarchy provides varying levels of interpretation that can be accessed based on the goals of the data analysis.

We describe a system called Subdue that discovers interesting substruc-

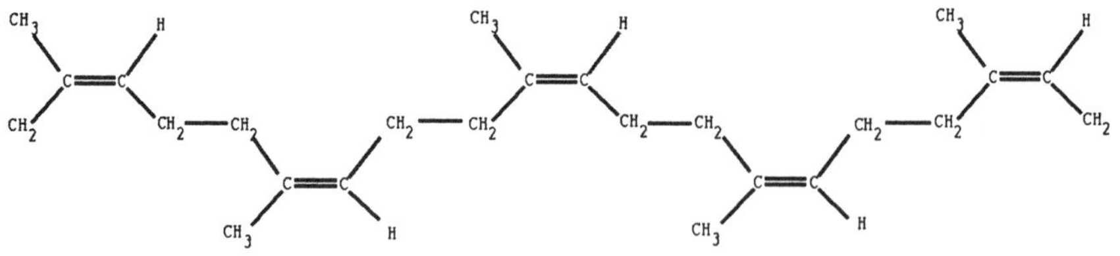

Figure 9.1   Natural rubber atomic structure.

tures in structural data based on the minimum description length (MDL) principle. The Subdue system discovers substructures that compress the original data and represent structural concepts in the data. By replacing previously discovered substructures, multiple passes of Subdue produce a hierarchical description of the structural regularities in the data. Subdue uses a computationally bounded inexact graph match that identifies similar, but not identical, instances of a substructure and finds an approximate measure of closeness of two substructures when under computational constraints. In addition to the MDL principle, other background knowledge can be used by Subdue to guide the search toward more appropriate substructures.

9.1   Algorithmic Techniques

The substructure discovery system represents structured data as a labeled graph. Objects in the data map to vertices or small subgraphs in the graph, and relationships among objects map to directed or undirected edges in the graph. A *substructure* is a connected subgraph within the graphical representation. This graphical representation serves as input to the substructure discovery system.

Figure 9.1 shows a sample graphical representation of an input file to Subdue that represents the atomic structure of natural rubber. Given this input database, Subdue discovers the subgraph shown in figure 9.2 as the best subgraph to describe the input database.

An *instance* of a substructure in an input graph is a set of vertices and edges from the input graph that match, isomorphically, to the graphical representation of the substructure. The input graph contains five instances of this concept in the database. As shown in figure 9.3, Subdue replaces the instances with a pointer to the concept definition and thus redefines the database in terms of this substructure. Figure 9.4 shows the substructure discovered by Subdue in another chemical compound, cortisone.

The substructure discovery algorithm used by Subdue is a computationally constrained beam search. The algorithm begins with the substructure

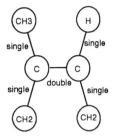

Figure 9.2   Discovered substructure.

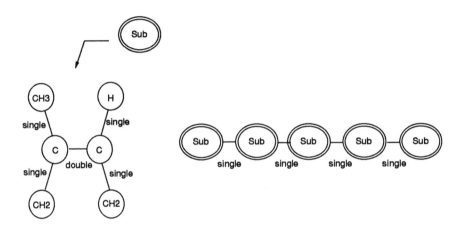

Figure 9.3   Compressed representation of the graph.

Original Database   Discovered Substructure

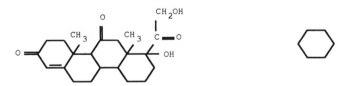

Figure 9.4   Cortisone atomic structure.

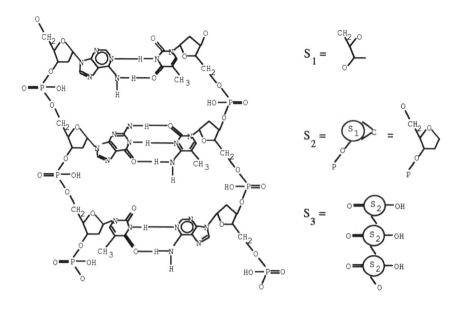

Figure 9.5   Sample results of Subdue on a DNA sequence.

matching a single vertex in the graph. Each iteration through the algorithm selects the best substructure and expands the instances of the substructure by one neighboring edge in all possible ways. The new, unique generated substructures (duplicate substructures are removed from the list) become candidates for further expansion. The algorithm searches for the best substructure until all possible substructures have been considered or the total amount of computation exceeds a given limit. The evaluation of each substructure is guided by the MDL principle and other background knowledge provided by the user.

Typically, once the description length afforded by an expanding substructure begins to increase, further expansion of the substructure will not yield a smaller description length. As a result, Subdue makes use of an optional pruning mechanism that eliminates substructure expansions from consideration when the description lengths for these expansions increase.

Figure 9.5 shows a sample input database containing a portion of a DNA sequence. In this case, atoms and small molecules in the sequence are represented with labeled vertices in the graph, and the single and double bonds between atoms are represented with labeled edges in the graph. Subdue discovers substructure S_1 from the input database. After compressing the original database using S_1, Subdue finds substructure S_2, which, when used to compress the database further, allows Subdue to find substructure S_3. Such repeated application of Subdue generates a hierarchical description of the structures in the database.

9.1.1 Substructure Discovery Using the MDL Principle

The *minimum description length* (MDL) principle introduced by Rissanen (1989) states that the best theory to describe a set of data is a theory that minimizes the description length of the entire dataset. The MDL principle has been used for decision tree induction (Quinlan and Rivest, 1989), image processing (Leclerc, 1989; Pentland, 1989), concept learning from relational data (Derthick, 1991), and learning models of nonhomogeneous engineering domains (Rao and Lu, 1992).

We demonstrate how the MDL principle can be used to discover substructures in complex data. In particular, a substructure is evaluated based on how well it can compress the entire dataset. We define the MDL of a graph to be the minimum number of bits necessary to describe the graph completely. Subdue searches for a substructure that minimizes $I(S)+I(G|S)$, where S is the discovered substructure, G is the input graph, $I(S)$ is the number of bits (description length) required to encode the discovered substructure, and $I(G|S)$ is the number of bits required to encode the input graph G with respect to S.

The graph connectivity can be represented by an adjacency matrix. Assuming that the decoder has a table of the unique labels in the original graph G, the encoding of the graph consists of the following steps:

1. Determine the number of bits *vbits* needed to encode the vertex labels of the graph.
2. Determine the number of bits *rbits* needed to encode the rows of the adjacency matrix A in which rows and columns represent graph vertices.
3. Determine the number of bits *ebits* needed to encode the edges represented by the entries $A[i,j] = 1$ of the adjacency matrix A.

The total encoding of the graph takes $(vbits + rbits + ebits)$ bits.

9.1.2 Inexact Graph Match

Although exact structure match can be used to find many interesting substructures, many of the most interesting substructures show up in a slightly different form throughout the data. These differences may be due to noise and distortion or may simply result from slight differences between instances of the same general class of structures.

Given an input graph and a set of defined substructures, we want to find those subgraphs of the input graph that most closely resemble the given substructures. Furthermore, we want to associate a distance measure between a pair of graphs consisting of a given substructure and a subgraph of the input graph. We adopt the approach to inexact graph match given by Bunke and Allermann (1983).

In this inexact match approach, each distortion of a graph is assigned a cost. A distortion is described in terms of basic transformations such as deletion, insertion, and substitution of vertices and edges. The distortion costs can be determined by the user to bias the match for or against particular types of distortions.

An inexact graph match between graphs g_1 and g_2 maps g_1 to g_2 such that g_2 is interpreted as a distorted version of g_1. Formally, an inexact graph match is a mapping $f\colon N_1 \to N_2 \cup \{\lambda\}$, where N_1 and N_2 are the sets of vertices of g_1 and g_2, respectively. A vertex $v \in N_1$ mapped to λ [i.e., $f(v) = \lambda$] is deleted. That is, it has no corresponding vertex in g_2.

Given a set of particular distortion costs as discussed above, we define the cost of an inexact graph match $cost(f)$ as the sum of the costs of the individual transformations resulting from f, and we define $matchcost(g_1, g_2)$ as the value of the least-cost function that maps graph g_1 onto graph g_2. If g_2 is isomorphic to g_1, the cost of the match will be 0. The cost of this match increases as the number of dissimilarities between the graphs increases.

Given g_1, g_2, and a set of distortion costs, the actual computation of $matchcost(g_1, g_2)$ can be determined using a tree search procedure. A state in the search tree corresponds to a partial match that maps a subset of the vertices of g_1 to a subset of the vertices in g_2. Initially, we start with an empty mapping at the root of the search tree. Expanding a state corresponds to adding a pair of vertices, one from g_1 and one from g_2, to the partial mapping constructed so far. A final state in the search tree is a match that maps all vertices of g_1 to g_2 or to λ such that for any edge between vertices in g_1, there is an edge between the corresponding vertices in g_2. The complete search tree of the example in figure 9.6 is shown in figure 9.7. For this example we assign a value of 1 to each distortion cost. The numbers in circles in this figure represent the cost of a state. As we are eventually interested in the mapping with minimum cost, each state in the search tree gets assigned the cost of the partial mapping that it represents. Thus, the goal state to be found by our tree search procedure is the final state with minimum cost among all final states. From figure 9.7 we conclude that the minimum-cost inexact graph match of g_1 and g_2 is given by the mapping $f(1) = 4$, $f(2) = 3$. The cost of this mapping is 3.

Given graphs g_1 with n vertices and g_2 with m vertices, $m \geq n$, the complexity of the full inexact graph match is $O(n^{m+1})$. Because this routine is used heavily throughout the discovery and evaluation process, the complexity of the algorithm can significantly degrade the performance of the system.

To improve the performance of the inexact graph match algorithm, we extend Bunke's approach by applying a branch-and-bound search to the tree. The cost from the root of the tree to a given node is computed as described above. Nodes are considered for pairings in order from the most

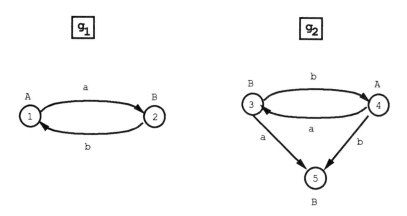

Figure 9.6 Two similar graphs g_1 and g_2.

heavily connected vertex to the least connected, as this constrains the remaining match. Because branch-and-bound search guarantees an optimal solution, the search ends as soon as the first complete mapping is found.

In addition, the user can place a limit on the number of search nodes considered by the branch-and-bound procedure (defined as a function of the size of the input graphs). Once the number of nodes expanded in the search tree reaches the defined limit, the search resorts to hill climbing using the cost of the mapping so far as the measure for choosing the best node at a given level. By defining such a limit, significant speedup can be realized at the expense of accuracy for the computed match cost.

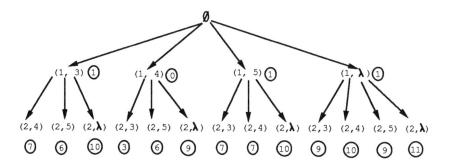

Figure 9.7 Search tree for computing $matchcost(g_1, g_2)$ from figure 9.6.

9.1.3　Adding Domain Knowledge to the Subdue System

The Subdue discovery system was initially developed using only domain-independent heuristics to evaluate potential substructures. As a result, some of the discovered substructures may not be useful and relevant to specific domains of interest. For instance, in a programming domain, the Begin and End statements may appear repetitively within a program; however, they do not perform any meaningful function on their own and so exhibit limited usefulness. Similarly, in a biochemical domain, some compounds or substructures may appear repetitively within the data; however, they may not represent meaningful or functional units for this domain. To ensure that Subdue's discovered substructures will be interesting and useful across a wide variety of domains, domain knowledge is added to guide the discovery process. Furthermore, using the domain knowledge can increase the chance of realizing greater graph compression compared with not using the domain knowledge.

In this subsection we present two types of domain knowledge that are used in the discovery process and explain how they bias discovery toward certain types of substructures.

Model Knowledge

Model knowledge provides the discovery system with specific types of structures that are likely to exist in a database and that are of particular interest to a scientist using the system. The model knowledge is organized in a hierarchy of and combinations of primitive well-recognized structures. This hierarchy is supplied by a domain expert.

Although the MDL principle still drives the discovery process, domain knowledge is used to input a bias toward certain types of substructures. First, the modified version of Subdue can be biased to look specifically for structures of the type specified in the model hierarchy. The discovery process begins with matching a single vertex in the input graph to primitive nodes of the model knowledge hierarchy. If the primitive nodes do not match the input vertices, the higher level nodes of the hierarchy are pursued. The models in the hierarchy pointed to by the matched model nodes in the input graph are selected as candidate models to be matched with the input substructure. Each iteration through the process, Subdue selects a substructure from the input graph that has the best match to one of the selected models and can be used to compress the input graph.

The match can be either a subgraph match or a whole graph match. If it is a subgraph match, Subdue expands the instances of the best substructure by one neighboring edge in all possible ways. The newly generated substructure becomes a candidate for the next iteration. If it is a whole graph match, the process has found the desired substructure, which is then used to compress the entire input graph. The process continues until either a substructure has been found or all possible substructures have been considered.

Representing an input graph using a discovered substructure from the model hierarchy involves additional overhead to replace the substructure's instances with a pointer to the model hierarchy. In some cases, a model definition also includes parameters that must be represented. After a substructure is discovered, each instance of the discovered substructure in the input graph is replaced by a pointer to a concept in the model hierarchy representing the substructure.

Graph Match Rules

At the heart of the Subdue system lies an inexact graph match algorithm that finds instances of a substructure definition. Since many of substructure instances can show up in a slightly different form throughout the data, and each of these differences is described in terms of basic transformations performed by the graph match, we can use graph match rules to assign to each transformation a cost based on the domain of usage. This type of domain-specific information is represented using if-then rules such as the following:

```
If (domain = Biochemical) and
    ((Vertex1 = HO and Vertex2 = OH) or
     (Vertex1 = OH and Vertex2 = HO))
Then (substitute-vertex-label cost = z)
```

The graph match rules allow us to specify the amount of acceptable generality between a substructure definition and its instances or between a model definition and its instances in the input graph. Given g_1, g_2, and a set of distortion costs, the actual calculation of similarity can be performed using the search procedure described above. As long as the similarity is within the user-defined threshold, the two graphs g_1 and g_2 are considered to be isomorphic.

We have tested the performance of Subdue with model knowledge, with graph match rules, and with no background knowledge. Performance is measured in terms of the amount of compression afforded by the highest valued discovered substructure and by the amount of substructure functionality. Functionality measures were provided by a team of eight scientists familiar with the application domain—functionality is measured on a scale of 1 to 5. When both types of background knowledge are used, the highest functionality ratings and compression result. The amount of compression resulting from the use of graph match rules alone is lower than using no background knowledge, and both of these cases result in lower human ratings than when both types of background knowledge are used. More details of these experiments are described in Djoko et al. (1996).

9.1.4   Scalability

A goal of knowledge discovery in database (KDD) systems is to discover knowledge in large databases that cannot be effectively processed by humans. For this reason KDD systems are required to handle very large databases. Unfortunately, most KDD systems are computationally expensive, and in most cases the resource requirements of a KDD system grow as the database becomes larger. For these reasons, researchers' attentions are shifting from developing new KDD systems to improving the scalability of existing KDD systems.

A goal of our research is to demonstrate that KDD systems in general, and Subdue in particular, can be made scalable by making efficient use of parallel and distributed hardware. One way to improve the scalability of a system is to make use of system resources to allow larger inputs and/or reduce run time. Clearly, increasing the resources of a single machine will not benefit us in the limit, as the input is usually large enough to exhaust the resources of a single machine. Thus, we have to make use of parallel algorithms. Porting a complicated system to a parallel environment involves consideration of the architecture as well as the nature of the problem itself.

A variety of methods can divide the computational effort among individual processors. However, many of these approaches store entire database on each contributing machine. Because we want Subdue to provide scalability in storage as well as computation, we introduce a data-partitioning approach, Static-Partitioning Subdue (SP-Subdue).

Using SP-Subdue we partition the input graph into n partitions for n processors. Each processor executes the sequential Subdue algorithm on its local graph partition and broadcasts its best substructures to the other processors. Each processor then evaluates the communicated substructures on its own local partition. Once all evaluations are complete, a master processor gathers the results and determines the global best discoveries.

The graph partitioning step is the most important step of the algorithm. The speedup achieved, as well as the quality of discovered substructures, depends on this step. In partitioning the graph we want to balance the work load equally among processors while retaining as much information as possible (edges along which the graph is partitioned may represent important information). SP-Subdue utilizes the Metis graph partitioning package (Karypis and Kumar, 1995). Metis accepts a graph with weights assigned to edges and vertices and tries to partition the graph so that the sum of the weights of the cut edges is minimized and the sum of vertex weights in each partition is roughly equal. We assign weights only to edges to ensure that each partition will contain roughly the same number of vertices. Edge weights are assigned so that the higher the frequency of occurrence of an edge's label, the higher the weight assigned to it. The idea behind this weight assignment is that the frequently occurring edge labels are more likely to represent useful knowledge. The run time of Metis to partition our

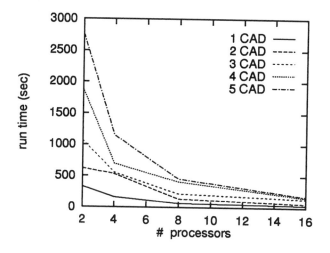

Figure 9.8 CAD database evaluation time.

test databases is very small (10 s on average) and is thus not included in the parallel run time.

Figures 9.8 and 9.9 graph the run time of SP-Subdue as the number of processors increases on a sample graph of two types. The first type of database is a CAD circuit description of an analog-to-digital converter provided by National Semiconductor (figure 9.8). The initial graph contains 8,441 vertices and 19,206 edges. In addition, we also tested an artificial graph (ART; figure 9.9) in which an arbitrary number of instances of a predefined substructure are embedded in the database surrounded by vertices and edges with random labels and connectivity. The tested artificial graph contains 1,000 vertices and 2,500 edges. To create larger databases with the same characteristics as the sample databases, we generated multiple copies of these graphs and merged the copies together by artificially connecting the individual graphs, yielding a new larger graph. Thus, in figures 9.8 and 9.9, "n CAD" refers to a database consisting of n copies of the original CAD database with joining edges added, and "n ART" refers to a database consisting of n copies of the artificial database with additional joining edges.

The speedup achieved with the ART database is always superlinear. This is because the run time of sequential Subdue is greater than linear with respect to the size of the database. Each processor essentially executes a serial version of Subdue on a small portion of the overall database, so the combined run time is less than that of serial Subdue. The speedup achieved with the CAD database is usually close to linear and is sometimes superlinear. Increasing the number of partitions always results in a better speedup.

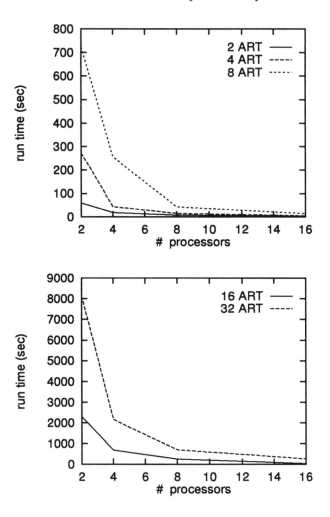

Figure 9.9   ART database evaluation time. Top, 2, 4, and 8 copies of the artificial database. Bottom, 16 and 32 copies of the artificial database.

Now we turn our attention to the quality of the substructures discovered by SP-Subdue. Tables 9.1 and 9.2 show the compression achieved for the ART and CAD databases when processed by a different number of processors. Regarding the ART database results (table 9.1), it is clear that the best compression achieved is the sequential version compression. This is expected because of the high regularity of the database. The ART database has one substructure embedded into it; thus, partitioning the database can only cause some instances of this substructure to be lost because of the edge cuts. As a result, we will generally not get better compression by partition-

Table 9.1 ART database compression results by number of processors.

Database	Number of processors				
	1	2	4	8	16
2 ART	.5639	0.543554	0.545256	0.564162	0.626758
4 ART	.5415	0.519963	0.321946	0.296317	0.225197
8 ART	.5418	0.518611	0.501251	0.310067	0.257190
16 ART	.5407	0.514947	0.504967	0.497538	0.303339
32 ART	.5187	0.507182	0.495180	0.492651	0.490307

ing, though some compression results improve due to the increased beam width provided by multiple processors. The compression achieved with a small number of partitions is very close to that of the sequential version.

Regarding the CAD database compression results (table 9.2), we find that a small number of partitions almost always results in a compression superior to that of the sequential version. The reason lies in the nature of the CAD database. As with many real-world databases, the CAD database contains many diverse substructures. Treating the entire database as a single partition will result in throwing away good substructures because of the limited search beam. When the database is partitioned among several processors, each processor will have a greater chance to deepen the search into the database because the number of embedded substructures is reduced, resulting in higher valued global substructures.

Because the data are partitioned among the processors, SP-Subdue can also utilize the increased memory resources of a network of workstations using communication software such as the Parallel Virtual Machine (PVM) system (Geist et al., 1994). We implemented SP-Subdue on a network of 14 Pentium PCs using PVM. The speedup results are given in tables 9.3 and 9.4.

By partitioning the database effectively, SP-Subdue proves to be a highly scalable system. SP-Subdue can handle huge databases when provided with the same amount of processing power and memory required by

Table 9.2 CAD database compression results by number of processors.

Database	Number of processors				
	1	2	4	8	16
1 CAD	.0746	0.156095	0.112458	0.028686	0.019010
2 CAD	.0746	0.082348	0.123302	0.034331	0.017073
3 CAD	.0746	0.060023	0.050129	0.033402	0.033586
4 CAD	.0746	0.101977	0.042417	0.048962	0.026821
5 CAD	.0746	0.079247	0.065364	0.032131	0.017944

Table 9.3   ART database
distributed version speedups.

	Number of PCs			
Database	2	4	8	14
2 ART	4.34	12.76	31.42	63.8
4 ART	2.08	15.69	56.56	139.33
8 ART	2.57	6.09	48.45	162.38
16 ART	2.87	9.96	20.53	171.09
32 ART	3.16	10.31	34.12	82.99

the sequential version to handle the same database, while discovering sub-structures of equal or better quality. One of our databases contains 2 million vertices and 5 million edges, yet SP-Subdue is able to process the database in less than 3 hours. The easy availability of SP-Subdue is greatly improved by porting the system to distributed systems. The minimal amount of communication and synchronization required makes SP-Subdue ideal for distributed environments. Using the portable message passing interface provided by PVM allows the system to run on heterogeneous networks.

9.2   Future Work and Generalizations

The increasing structural component of today's biological databases requires data mining algorithms capable of handling structural information. The Subdue system is designed specifically to discover knowledge in structural databases.

We have described a method for integrating domain-independent and domain-dependent substructure discovery based on the minimum description length (MDL) principle. The method is generally applicable to many structural databases, such as chemical compound data, computer-aided design data, computer programs, and protein data. This integration improves Subdue's ability both to compress an input graph and to discover substructures relevant to the domain of study. In addition, we show that making

Table 9.4   CAD database
distributed version speedups.

	Number of PCs			
Database	2	4	8	14
1 CAD	1.48	3.37	7.99	18.73
2 CAD	1.88	3.52	13.64	28.69
3 CAD	1.98	5.54	14.76	33.79
4 CAD	2.58	8.81	15.81	47.18
5 CAD	2.13	6.25	23.27	52.58

efficient use of parallel and distributed resources can significantly improve the run-time performance of data-intensive and compute-intensive discovery program such as Subdue. We are currently evaluating linear discovery algorithms for inclusion in our system and are adapting our distributed algorithms for application to additional discovery systems.

Chapter 10

Overview: A System for Tracking and Managing the Results from Sequence Comparison Programs

David P. Yee, Judith Bayard Cushing,
Tim Hunkapiller, Elizabeth Kutter,
Justin Laird, and Frank Zucker

The Human Genome Project was launched in the early 1990s to map, sequence, and study the function of genomes derived from humans and a number of model organisms such as mouse, rat, fruit fly, worm, yeast, and *Escherichia coli*. This ambitious project was made possible by advances in high-speed DNA sequencing technology (Hunkapiller et al., 1991). To date, the Human Genome Project and other large-scale sequencing projects have been enormously successful. The complete genomes of several microbes (such as *Hemophilus influenzae* Rd, *Mycoplasma genitalium*, and *Methanococcus jannaschii*) have been completely sequenced. The genome of bacteriophage T4 is complete, and the 4.6-megabase sequence of *E. coli* and the 13-megabase genome of *Saccharomyces cerevisiae* have just recently also been completed. There are 71 megabases of the nematode *Caenorhabditis elegans* available. Six megabases of mouse and 60 megabases of human genomic sequence have been finished, which represent 0.2% and 2% of their respective genomes. Finally, more than 1 million expressed sequence tags derived from human and mouse complementary DNA expression libraries are publicly available. These public data, in addition to private and proprietary DNA sequence databases, represent an enormous information-processing challenge and data-mining opportunity.

The need for common interfaces and query languages to access heterogeneous sequence databases is well documented, and several good systems are well underway to provide those interfaces (Woodsmall and Benson, 1993; Marr, 1996). Our own work on database and program interoperability in this domain and in computational chemistry (Cushing, 1995) has shown, however, that providing the interface is but the first step toward making these databases fully useful to the researcher. (Here, the term "database" means a collection of data in electronic form, which may not necessarily be physically deposited in a database management system [DBMS]. A scientist's database could thus be a collection of flat files, where the term "database" means "data stored in a DBMS" is clear from the context.)

Deciphering the genomes of sequenced organisms falls into the realm of analysis; there is now plenty of sequence data. The most common form of sequence analysis involves the identification of homologous relationships among similar sequences. Two sequences are homologous if they share a common ancestor sequence that has diverged over time. There are multiple types of homology. Orthologs refer to homologous genes that occur in different species; the divergence of the gene sequences is due to speciation, and the function of the orthologs is generally conserved. Paralogs are genes that arose by duplication and subsequent divergence of a gene within a genome. The redundancy generated by duplicating a gene within a genome allows a paralog to evolve a new (yet related) function.

A common way of analyzing sequence databases is to use programs that compare one sequence against a database of other sequences. These programs estimate the similarity between two sequences by identifying substrings derived from each sequence that, when aligned, minimize the cost of converting one substring into another. The cost of substituting one letter for another in a sequence is given by a substitution matrix (Needleman and Wunsch, 1970; Dayhoff et al., 1978; Henikoff and Henikoff, 1992). The problem of finding the best local alignment of two sequences that maximize similarity (and minimize cost) can be solved rigorously using dynamic programming (Smith and Waterman, 1981a). Many other algorithms (Pearson and Lipman, 1988; Altschul et al., 1990) solve a similar problem using heuristics to speed up the process at the cost of not being guaranteed the optimal solution.

By comparing a new, uncharacterized sequence against a database of known, characterized sequences, the uncharacterized sequence can be associated with the most similar known sequence. If the degree of similarity is strong, then the two sequences may share a homologous relationship, and the new sequence may be assigned potential biological functions that can be tested in the laboratory or classified into a functional family. All sequence comparison methods, however, suffer from certain limitations. For example, there are no deep principles from which one can derive a perfect, comprehensive substitution matrix. Also, the problem of determining where to place insertions and deletions between two aligned sequences and at what cost is far from solved. Consequently, a researcher may want to try several different matrices and several different gap models during the course of a single study. For any sequence, dozens or even hundreds of searches may be performed.

The system described in this chapter integrates and tracks inputs and results from sequence comparison programs. It provides the capability of organizing these results into "clusters" that can be marked, named, annotated, and manipulated. Objects in clusters can themselves be sorted, filtered, and annotated. A prototype of the system, which we call Overview, is implemented in Smalltalk.

10.1 Related Work

Research in data mining, scientific databases, distributed computing, and computational molecular biology is all relevant to our project. "Knowledge discovery in databases and data mining aim at semiautomatic tools for the analysis of large datasets" (Mannila, 1996). While our project aims to provide a tool that molecular biologists can use to gain knowledge from large collections of data, namely, the genomic databases currently published electronically, the system we are developing grew out of our previous work to improve the interoperability of distributed scientific applications (Cushing, 1995), rather than the hypothesis that molecular biologists could "mine" existing datasets. We determined, first, that using computational applications was primarily the purview of a small, elite group of scientists and that examples of past computations would help nonspecialists gain access to this important class of applications. Second, we discovered that many computational scientists, molecular biologists included, needed help in keeping track of input and output files. Third, we reasoned that the only way to accomplish these two objectives would be to provide the scientist with a common interface to the many distributed tools available.

Since we started our work in 1991, much progress has been made in improving the interoperability of programs and databases. Technological advances such as Microsoft's object linking and embedding (OLE; OLE Team, 1996), AppleEvents (Apple Computer, 1997), ToolTalk (SunSoft, 1993), the Common Object Request Broker Architecture, and CORBA-compliant implementations such as ParcPlace's Distributed SmallTalk (ParcPlace, 1997) are beginning to make interoperability technically feasible.

Regarding scientific databases and applications, mainstream work in distributed architectures, such as that of Wiederhold and his collaborators (see Wiederhold, 1992), is highly relevant, as it provides a common understanding of the problem and a language and context for discussing the alternative solutions. To make this distributed technology work for the scientist, however, agreement as to the meaning of the data being exported and imported is required (Cushing et al., 1993). There has been considerable interest of late in using advanced data models for integrating scientific data and letting applications interoperate with one or more models through a uniform interface. Rieche and Dittrich (1994) and Kemp et al. (1996) present two examples of approaches for integrating molecular biology data, using object-oriented and functional models, respectively. Object-oriented technology has also been used by others to provide more intelligent interfaces to scientific applications, such as the SCENE system of Peskin and Walther (1994) and the experiment management system of Sparr et al. (1991). The ZOO system of Ioannidis, Livny, and others (Wiener and Naughton, 1994; Ioannidis et al., 1996) is probably the most similar work to our own, in that it is an object-oriented system that focuses on experiment management.

Within the mainstream of database research, three areas deserve special attention. (1) Both semantic and object-oriented data modeling are important because of the importance of representing complex scientific objects independently of any physical schema (Brachman and Schmolze, 1985; Borgida et al., 1989; Atkinson et al., 1990; Manola and Dayal, 1990; Agrawal et al., 1996). (2) Research on personal databases and laboratory notebooks is relevant because of the interplay between an individual scientist's private databases and laboratorywide or public databases (Weissman, 1989). (3) Finally, the ultimate success of a system such as Overview hinges on how effectively the user interface presents the information. Ben Shneiderman's work is particularly relevant to these efforts (see Shneiderman, 1993).

10.2 Molecular Biology Applications and Databases

Here we briefly describe the databases, database interfaces, and programs commonly available to molecular biologists. Molecular biology is one of the major areas of modern biology, with its beginning typically given as 1953 when Watson and Crick proposed the double helical structure of DNA. Introductory texts in molecular biology and genetics abound, and some tutorials are written specifically for mathematicians or computer scientists (Frenkel, 1991; Smith, 1994; Lander and Waterman, 1995; Waterman, 1995). For the purpose of understanding this chapter, it is sufficient to know that a DNA sequence is a sequence of letters representing nucleotides and that a DNA string can be translated to another string of letters representing amino acids, the letters of a protein sequence. The search for genes involves finding the sequences of DNA from specific organisms that code for certain proteins with physical properties of interest, or finding functions for the proteins potentially encoded by uncharacterized "open reading frames" (ORFs), stretches of DNA that appear likely to encode some functional gene product.

Molecular biology provides one of the most challenging application domains for database research (Kemp et al., 1996). It also provides one of the most productive, since all major genetic science publications now require researchers to publish their sequences electronically, and the Human Genome Project has set aside a considerable percentage of funding for bioinformatics research. As a result, (1) much data is available electronically and available to researchers, who very much want to use this data effectively, and (2) much research and development have been done to provide common file formats, databases, interfaces to databases, and computational applications.

Molecular biologists are now using private and public databases, perhaps because these data are (at least on the surface) easily represented as ASCII character strings (Burks, 1989; Fickett and Burks, 1989; Computer Science and Technology Board, 1990; Lander et al., 1991). Between 1985 and 1989 the number of nucleotides in centralized DNA databases increased sevenfold, from 3 to 21 million (Waterman, 1989). Automated sequencing methods promise to increase this database 10-fold every three years; the 1994 goal was to sequence 160 million base pairs per year (up from 16 million in 1989; Lander et al., 1991), a goal that was exceeded—over 160 million bases were added to GenBank alone in 1995. Some researchers contend that data are currently accumulating at near exponential rates. GenBank, the Human Genome Project (Letovsky et al., 1990), and the Brookhaven Protein Database (Bernstein et al., 1977; Abola et al., 1987) are perhaps the most ambitious of these public database projects, but there are a number of other genome-related databases (Courteau, 1991). As of December 1997, GenBank contains approximately 746 million bases of genomic sequence and 512 million bases of "expressed sequence tags" (ESTs), which code for 85 million amino acid residues in GenPept. Swiss-Prot has 22 million amino acids; the Protein Information Resource (PIR) contains 32 million amino acids. The sequences in these databases, in addition to the annotations included for each sequence, account for several gigabytes of data.

Research and development in molecular bioinformatics can be categorized as follows:

1. *Publicly available sequence databases.* These provide remote viewing and access, and query for large numbers of users, not all of whom are researchers. These databases usually provide interfaces for researchers to enter new sequences and for a curator or database administrator to validate those entries. Molecular biology databases are supported by national agencies and include those for both DNA and proteins, such as GenBank (Benson et al., 1996) and Swiss-Prot (Rodriguez-Tome et al., 1996). Molecular biologists and computer scientists are currently working to provide easy-to-use interfaces to these databases, as well as effective means for researchers to update them.

2. *Heterogeneous molecular biology databases.* The ability to write queries across the public databases is highly desirable; researchers would like to treat the necessarily heterogeneous databases as a single federated database. Interoperability of heterogeneous databases is an important area of research (Rieche and Dittrich, 1994; Buneman et al., 1995; Karp et al., 1996; Kemp et al., 1996).

3. *Workstations for molecular biologists.* As computers are integrated into the work of molecular biologists for sequencing DNA and proteins, these researchers would like to use a single application to access a number of databases, computational programs, and laboratory results. The Genome Topographer (GT) project, formerly at Cold Spring Harbor (Marr, 1996), is an example of such a workstation.

4. *Toolkits and new languages.* Specific tools, such as the Object Protocol Model (OPM), assist in these efforts and may be relevant to database problems in other scientific domains. The work of Buneman and Davidson (see Buneman et al., 1995) applies new functional language paradigms to problems faced by developers of molecular biology database applications.

5. *Computational tools to support molecular biologists.* The most prevalent class of computational biology applications compare a subject sequence against many object sequences. The programs of primary interest to us include the various versions of BLAST (Basic Local Alignment Search Tool; Altschul et al., 1990) and Fasta (Pearson and Lipman, 1988) and those that implement the Smith-Waterman algorithm (Smith and Waterman, 1981b). BLAST tries to find patches of regional similarity, rather than trying to find the best alignment between a query and an entire database sequence; alignments generated with older versions of BLAST do not contain gaps. Fasta also tries to find patches of regional similarity; alignments generated with Fasta can contain gaps. Most molecular biologists use the implementations of BLAST and Fasta that are available on the Internet, for example, at the National Library of Medicine and National Institutes of Health/National Center for Biotechnology Information (NCBI) Web site (Woodsmall and Benson, 1993). The Smith-Waterman algorithm computes the mathematically optimal alignment and similarity (or matching score) of a query sequence compared with each sequence record in the database; implementations of this algorithm abound, and most laboratories use local rather than Web-based programs.

Other types of computational biology programs that are of importance include applications that perform motif searches and profile searches. Hidden Markov models are also used to perform database searches. CRI-MAP (Lander and Green, 1987) is used for chromosome mapping and has options that can generate maps, calculate order probabilities, and print recombination data.

6. *Protein and crystallographic information and graphical viewing,* such as the Macromolecular Crystallographic Interchange Format (mmCIF; Pu et al., 1992) and its corresponding toolbox. mmCIF provides an extension of the format for exchanging crystallographic information for large molecules; it also provides a publically available data format for visualization and computational tools.

7. *Information and knowledge-based systems for the study of systems at the cellular level,* such as Karp's EcoCyc (see Karp et al., 1996). EcoCyc provides a tool for examining metabolic pathways; its authors aim to eventually relate metabolic data with genetic information.

Overview is most closely related to systems that "intelligently" search heterogeneous databases and workstations for molecular biologists, but it differs in that we focus on one aspect of the ideal workstation: integrating and reasoning about results of sequence comparisons from different programs. The system is designed to either function independently or be incorporated into a full-service workstation such as Genome Topographer.

10.3   Algorithmic Techniques

10.3.1   The Overview Project

Our work in database and program interoperability indicated to us that, once interoperability is achieved, scientists will be faced with a new problem: integrating, organizing, and reasoning about many more computational application results than they now have. The Extensible Computational Chemistry Environment (ECCE) at the Environmental Molecular Sciences Laboratory at the Department of Energy's Battelle Pacific Northwest National Laboratory in Richland, Washington, has endeavored to solve this problem by implementing a spreadsheetlike capability so that researchers can simultaneously view summary results of many computational experiments (D. Feller, personal communication). User evaluation of this idea has been enthusiastic, although ECCE is not yet installed as a production system.

The problem of integrating, organizing, and reasoning about computational application results is particularly critical in molecular biology applications where a major researcher activity involves the comparison of a subject sequence to many target sequences. The programs used to perform sequence comparisons typically generate a series of scores that estimate the degree of similarity between a subject sequence and many target sequences. Some comparison algorithms (such as BLAST and Fasta) also generate an estimate of the probability that the relationship between two sequences could have been observed by chance. A very low probability indicates that the relationship between two sequences is highly significant, whereas a relatively high probability suggests that the relationship may be due to random chance. Many of the most interesting relationships between sequences occur in the "twilight zone," where the significance of the relationship between two sequences is not clearly resolved by a particular algorithm (i.e., the probability or "pval" is high). Consequently, likely candidates must be analyzed manually by a knowledgeable expert in order to separate out the true, interesting relationships from the noise.

Figure 10.1 shows schematically how the output from a single comparison is often split apart by researchers (conceptually if not in actuality) so that items from different comparison searches can be compared with each other. In this figure, A refers to the subject sequence, Bi to target sequences, and F1 and G1 to invocations of programs F and G, respectively. Ri and Rri indicate ratings that programs F and G associate with

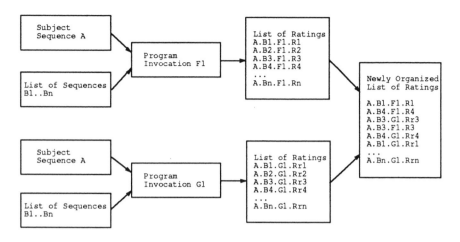

Figure 10.1 Main idea of Overview.

the comparisons between sequences A and Bi. The researcher wants to move search result elements from one place to another, reorder them, and maintain metadata about the biological data used in the search and about the invocation of the comparison programs such as parameters. Metadata are not shown in the figure.

The goal of Overview is to provide the researcher with more information than is available from applying a single program to a subject sequence and a database of object sequences to find novel, interesting relationships among a set of sequences. Overview is intended as a tool for storing, organizing, and analyzing the output of various genetic sequence comparison programs. It does not (yet) provide a mechanism for accessing genetic sequence databases or for directly invoking genetic sequence comparison programs. Our intention is to provide some of these latter facilities in only a rudimentary way and eventually to provide an interface between Overview and one or more "full-service" molecular biology workstations such as Genome Topographer.

10.3.2   Overview Functional Requirements

Our first objective was to produce a prototype database and browser that would keep track of sequences, comparison "runs," and search result elements. Presently, sequence comparison searches are performed externally to Overview, and then the search results are loaded into the system using one of the parsers built into Overview. The current system is a data repository and browser for computational experiment results and metadata. The browser functionality enables the researcher to store and access both the

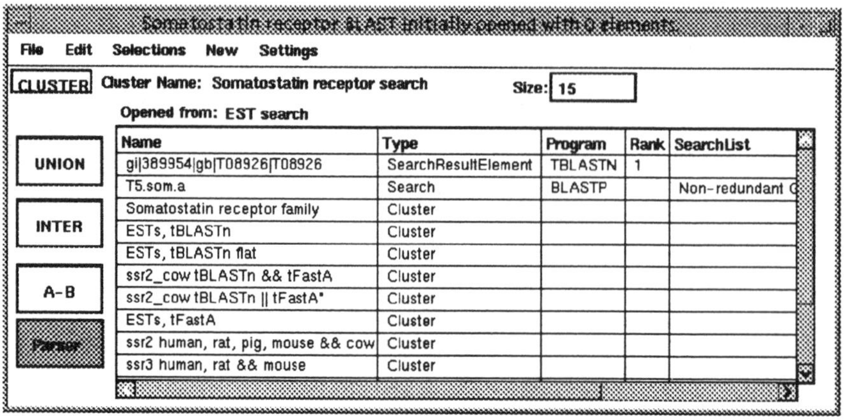

Figure 10.2   Heterogeneous cluster: related objects of different types.

parameterizations used for a series of comparison "runs" and the results generated by each run. Most importantly, Overview provides a mechanism to combine, order, and filter the results of these comparisons to facilitate analysis and discovery.

Overview has six major functions:

1. *Parse sequence search information.* Output files generated by BLAST, Fasta, and the Smith-Waterman programs can be parsed into the system. The system will also recognize output files from other comparison programs as well as sequence database formats such as GenBank, ASN.1, and automatic sequencer formats such as ABIF or SCF. Once parsed, comparison ratings are stored in Overview. Subject and target sequences are stored as references to the databases or local files from which they were derived. Overview maintains references to the output files in such a way that more detailed information can be retrieved and displayed.

2. *Organize sequence search information.* This involves creating containers that can group Overview objects of any type (figure 10.2). These containers are called "clusters." Clusters can be generated both manually by a user or automatically. Parsing a search output file into Overview, for example, automatically places the data into a cluster (see figure 10.3). Clusters contain several types of information:

   (a) DNA or protein sequence references, for retrieval of sequence data that are stored locally or remotely;
   (b) properties that can be associated with sequences, genes, and gene products;

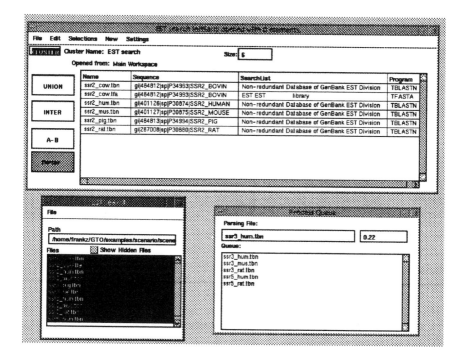

Figure 10.3   Search results being parsed into an Overview cluster. The file browser (lower left) allows users to select, view, and parse search result files into an Overview cluster (top). In this example, the parser is working on the seventh of 11 selected TBLASTN and TFasta files. The process queue (lower right) shows the selected files still to be parsed.

(c) information about database searches, including the program used to perform a search, the parameters with which the program was invoked, the location of the raw search results, the sequence database that was searched, and a list of search results elements, where the order of the elements is user defined;

(d) user profiles and information about scientific projects; and

(e) other clusters.

Some of the information within clusters is entered manually by the researcher, but sequence search result files and protein and DNA sequence files can be loaded into the system automatically.

3. *Analyze data.* Once sequence search information has been placed into a cluster, objects within the cluster can be displayed in sorted, filtered, and user defined order (see figures 10.4 and 10.5). A method is provided for searching the data and then grouping the results of comparison searches into new clusters. Data analysis is also facilitated by allowing researchers to annotate and name the objects they store

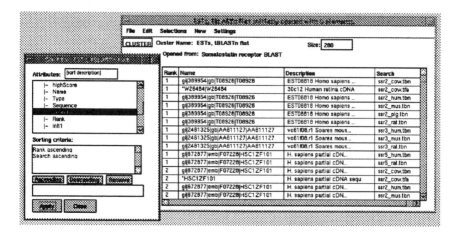

Figure 10.4 Automatic sorting: users can select fields and direction for sorting a cluster. A set of related searches placed in a single cluster can be "flattened" to replace each search with all of its SearchResult elements (e.g., the top 25 matches for each of 11 searches). This cluster can then be sorted by Rank and Search to put the best matches from each search at the top. In this example, one sequence (EST06818) is ranked first in 5 of 11 searches.

in the system. Descriptive information about sequences helps characterize both subject and target sequences. In sum, researchers will be able to organize data into clusters, order their contents, and display the cluster elements in a spreadsheet-style format. In the future, some specific functions or algorithms for comparing sequence search results will be implemented.

4. *Store data persistently.* To facilitate persistent storage, backup, and the sharing of data, Overview writes stored objects to flat files. In the future, a Gemstone or other database connection may be implemented for this purpose.

5. *Access remote databases.* Overview should be able to retrieve information from various sequence databases. For now, we retrieve data from the NCBI, where most of the major sequence databases are available.

6. *Launch searches.* Overview will eventually invoke local sequence comparison programs as well as automatically parse in the results generated by these programs.

10.4   The Overview Information Model

Our current information model reflects a preliminary requirements analysis of the molecular biologists' information needs for tracking the results of

Sequence	Description	Program	Rank	pN	n	reading Frame	high Score	z-sc	init1	initn	E
gi|389954|gb|T08926|T08926	EST06818 Homo sapiens ...	TBLASTN	1	1.0d-35	2	+1	236				
T08926	EST06818 Homo sapiens cDNA c	TFASTA	3				308	441.1	300	300	1.2e-18
gi|672877|emb|F07226|HSC1ZF101	H. sapiens partial cDN..	TBLASTN	2	2.1d-21	3	+1	158				
*HSC1ZF101	H. sapiens partial cDNA sequ	TFASTA	2				318	458.2	154	229	1.7e-17
*W26484|W26484	30c12 Human retina cDNA	TFASTA	1				330	466.2	214	294	4.7e-18
gi|1307183|gb|W26484|W26484	30c12 Human retina cDN..	TBLASTN	4	6.0d-19	3	-1	158				
gi|2461325|gb|AA611127|AA611112	vo61f08.r1 Soares mous...	TBLASTN	3	4.7d-20	4	+1	150				
emb|AA611127|AA611127	vo61f08.r1 Soares	TFASTA	6				249	350.9	177	213	1.2e-11
*N93967	za66d09.r1 Homo sapiens cDNA	TFASTA	4				260	371.1	244	244	9.3e-13
gi|1286298|gb|N93967|N93967	za86d09.r1 Homo sapien...	TBLASTN	5	9.8d-18	1	+2	190				
gi|1912557|gb|AA274112|AA274411	vb92d04.r1 Soares mous...	TBLASTN	21	1.6d-9	2	+3	107				
emb|AA274112|AA274112	vb92d04.r1 Soares	TFASTA	5				248	351.0	129	183	1.2e-11
gi|1277557|gb|W04836|W04836	za81f05.r1 Soares feta...	TBLASTN	6	1.4d-14	2	+1	167				
*W04836|W04836	za81f05.r1 Soares fetal	TFASTA	12				215	304.7	200	200	4.7e-9
gi|1182659|gb|H97311|H97311	EST42j10 WATM1 Homo sa...	TBLASTN	6	3.5d-14	2	+2	122				
*H97311	EST42j10 WATM1 Homo sapiens	TFASTA	7				242	343.6	150	191	3.1e-11
*emb|AA203362|AA203362	zx54h11.r1 Soare	TFASTA	17				208	289.9	95	132	3.1e-8

Figure 10.5 User-defined contents and order: SearchResult elements from two different search algorithms placed into one cluster for comparison. In this example, sequences found near the top in both TBLASTN and TFasta searches are displayed in pairs. Note that neither the sequence names nor descriptions match exactly, due to differences in databases used and text truncation by the search engines.

molecular biology experiments. Independent of any particular implementation, this information model forms a basis for the logical and physical system design.

Figure 10.6 is a graphical representation of our model. Entities (objects) are represented as boxes and relationships between entities as lines between boxes. One-to-one relationships are represented as unadorned lines, and one-to-many relationships as lines with a solid circle at the "many" end. Many-to-many relationships have two solid circles, one at each end. Optional 1–2 or 1–many relationships are indicated with an open circle at the 1-end.

For simplicity of illustration, we have divided our conceptual model into two parts: the Domain-Specific Information Model, which contains all domain objects, and the other model, which contains information about users, projects, and clusters. All objects in the conceptual model are to be implemented as persistent.

For the purposes of this chapter, we have abbreviated the discussion to the main entities of interest to molecular biologists. The full model is available in Cushing et al. (1997).

10.4.1 Sequences, Genes, and GeneProducts

A *sequence* is a string made up of an alphabet of either four nucleotides (symbolically represented as A, C, G, or T) if DNA or 20 amino acids if protein. A sequence usually encodes one or more gene products and is one of the primary objects of interest. (Some DNA sequences are "noncoding" in

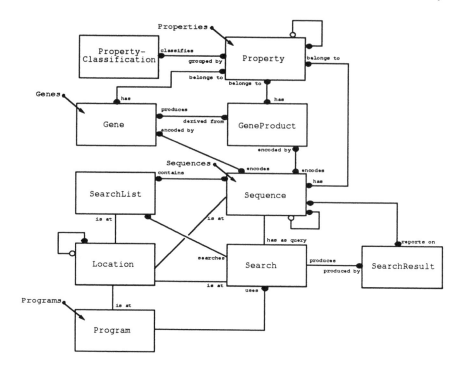

Figure 10.6   Overview information model.

the sense that they are not translated into an actual gene product.) The actual DNA or protein sequences are not permanently stored in Overview. Instead, pointers to their sequence locations (i.e., source databases and sequence accession numbers) allow them to be retrieved when needed.

Sequences are the major inputs to search programs and can be the subject or the object of a search. To distinguish search subject and object, we call the subject a *probe*. A probe can be a nucleic acid or protein sequence, a motif (Bairoch, 1993), or a profile (Gribskov et al., 1987). Because subject sequences are so ubiquitously used, they may be cached in Overview. A probe (or subject sequence) may be composed of subprobes (or subsequences). For example, a probe may be defined as a set of three different motifs. The order of such subsequences may or may not matter depending on the specific comparison program used.

Sequence attributes include name, a unique identifier from the probe's source repository (e.g., its accession number), the location of the source repository, *ProbeType*, and *SubProbe*.

A *gene* is defined as either a unit of genetic information that encodes one trait, or a piece of DNA encoding a protein or functional RNA. A gene is the most general level of biological organization in Overview; each biological entity in the database can potentially be traced back to a single gene.

A *GeneProduct* specifies the physical manifestation of a specific gene by collecting sequences or probes into a named entity. The sequences related to GeneProduct are typically probes.

10.4.2   Properties and PropertyClassifications

Properties are used to organize sequences into similar groups. A property is a biological characteristic such as "cancer causing," "produces hemoglobin," or "hydrophobic" and is used to justify the classification and organization of sequences into groups. Any number of properties can be associated with a sequence and vice versa. *PropertyClassifications* group similar properties together. Examples of property classes are "motif," "percent sequence identity," "function," "sequence family," and "structure family."

10.4.3   Search, SearchList, and SearchResult Element

A *search* is the invocation of a sequence comparison program. When researchers want to find more information about a particular sequence, they will run a program that compares the probe (subject sequence) against a collection of other (target) sequences. A *SearchList* is the collection of target sequences against which the probe is compared.

A search yields many SearchResult elements, typically one for each element of the SearchList that compares "favorably" with the probe. Each SearchResult element shows the similarity between the probe and one sequence from the SearchList.

10.4.4   SequenceRepository and SequenceDB (SeqDB)

A *SequenceRepository* is a local file (or database) containing many sequences of interest. A *SequenceDB* (or SeqDB) is a large database of released genomic information, such as GenBank or Swiss-Prot.

10.4.5   Program, Location, Host, and Archive

A *program* refers to an application program that can compare a probe to all the elements in a SearchList. A *location* is used to specify the physical location of data or programs referenced by the system, usually a file or directory that can be accessed by a host computer. A *host* is a computer on which a database or program resides; it is another type of location. An *archive* is a long-term storage location.

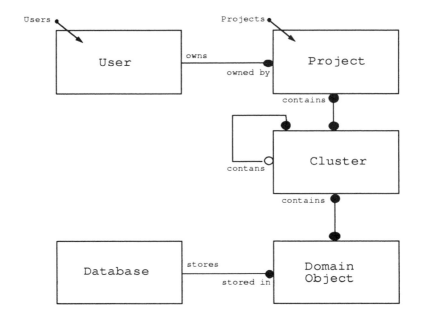

Figure 10.7   Users, projects, clusters, and database objects.

10.4.6   Users, Projects, and Clusters

The Overview information model includes a structure for identifying data about molecular biologists whose experiments are in the database, as well as data about their projects. Data about projects are organized into clusters (see figure 10.7). A *user* is a researcher who uses Overview. *Projects* allow the researcher to group together the information in Overview according to research project. Each project contains one or more clusters.

Originally designed to group probes with similar properties, *clusters* can be used to organize any domain object (including other clusters). A cluster is, in effect, an untyped collection that is controlled directly by the user and can be viewed in many ways. Clusters are flexible enough to do a wide variety of tasks but are mostly used like lists—to identify and select items of interest. Researchers are typically interested in a relatively small number of primary attributes of the objects in a cluster, such as name, data type, and a unique identifier (e.g., accession number) if it differs from the object's name; these attributes are displayed automatically. Researchers will be able to view on demand all other attributes of the elements within a cluster. While the types of cluster elements will usually be homogeneous, or grouped by type, it is critical to provide ways of including heterogeneous types in a single cluster. This heterogeneity complicates considerably the layout of the cluster display.

The entities within a cluster can be sorted, filtered, and explicitly selected for the purpose of viewing, browsing, and exporting to other clusters. The objects in a cluster are presented as a list. Cluster behavior is defined in detail as follows:

1. *Sorting* (figure 10.4) defines the order in which the domain objects within a cluster are presented to the user. Each object in a cluster carries a description of the object's possible sort criteria. The user can choose a sort criterion from a list of all possible sort criteria defined for any of the objects within the cluster. If a sort criterion is selected that is not defined for all the members of the cluster, those objects for which the criterion is not defined can be automatically filtered out or included at the end of the sorted list of cluster elements.

2. *Filtering* is used to limit the list of displayed objects in a cluster. The user can include or exclude specific object types. For example, one could limit the list of displayed objects to only include probes. Multiple filtering conditions can be specified and combined using And, Or, and Not operations.

3. *Constraining*, like filtering, is used to limit the objects associated with a cluster. Whereas filtering affects the display of cluster objects, constraining limits the objects that can be in a cluster. For example, the user can constrain a cluster to contain only genes or only objects that have definitions for specific data fields. For example, a cluster named "BlastSearches" could be defined to contain only those search instances where the program used was a BLAST variant.

4. *Selecting* enables the user to explicitly select specific objects in the displayed cluster. In addition, simple commands such as "select all," "unselect all," and "invert" are provided.

5. *Undo* provides a mechanism to undo the last filter, sort, or select operation.

6. The user can also *export* selected items to a new cluster; *save* the current combination of filters, constraints, and selections; and *print* the contents of a cluster as text. *Viewing* allows the user to take a more detailed look at a particular object, that is, to see all associated attributes.

7. Overview also supports *multiple cluster* operations to combine clusters using intersect, union, and difference operations (see figure 10.8). These operations provide the ability to determine which objects belong to multiple clusters and to combine data residing in multiple clusters into a single cluster. SearchResult elements from two clusters can also be combined via an interleaving process, such that the order of one cluster remains constant and the synonymous objects from the second cluster are interleaved with the corresponding elements from the first cluster.

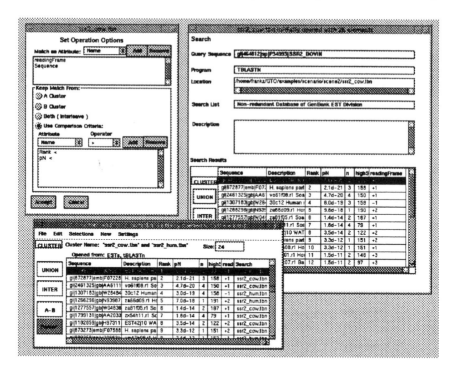

Figure 10.8 Multiple cluster operations on searches: user-defined intersections, unions, and differences. Users can select which attributes to compare and which elements to keep when performing set operations with clusters. In this example, we take the intersection of the SearchResult element clusters of two TBLASTN searches. TBLASTN can match a protein sequence to any one of six possible reading frames in a DNA sequence. Therefore, in addition to matching on sequence name (the default), here we also require that reading frames match, to eliminate the (very rare) cases where searches find the same sequence in different reading frames. We also set the comparison criteria for which element to keep in the resulting cluster: take whichever has the lower rank, and if the ranks match, take the one with the lowest pN.

10.5   Biological Experiments and Applications: User Scenarios

Here we give some scenarios describing how a researcher might use Overview. Three classes of problems are described: (1) searching for new protein family members in a database of novel sequences, (2) characterizing a set of novel sequences, and (3) analyzing a genomic DNA fragment.

10.5.1   Discovery of Novel Protein Family Homologs

The availability of novel sequence data in the form of expressed sequence tags (ESTs) gives researchers an opportunity to identify expressed genes that are homologous to known protein families. ESTs are relatively short sequences (~300 base pairs) that "tag" genes actively transcribed in cells. Since an average gene product (mRNA) is about 2,000 base pairs in length, any given EST represents only a fraction of the total gene. By assembling overlapping ESTs (Sutton et al., 1995), one can generate a longer sequence that represents the gene from which the contig ESTs were derived. The first challenge, however, is to identify a set of interesting ESTs. One approach is to characterize a protein family by collecting all known family members. Given a starting sequence of a known family member, the program BLASTP can be used to compare the sequence against a SearchList such as a nonredundant protein database (Bleasby and Wootton, 1990). The BLASTP output can be parsed into Overview and loaded into a cluster. By sorting the SearchResult elements by score (or pval) and filtering out scores that fall below a threshold, one can define a sequence set of all known family members.

Once a cluster of protein family members is generated, each protein sequence can be compared against another SearchList consisting of EST sequences using a variety of sequence comparison programs such as TBLASTN, Fastx, and Smith-Waterman. Each of the output files can be loaded into a cluster, ordered by score, and filtered with a score threshold. By combining all the SearchResult clusters using union operations, one can generate a master cluster containing a list of potentially interesting ESTs (see figure 10.9). For each EST in the master EST cluster, an EST assembly program (e.g., TIGR assembler) can be applied to define a cluster of contigs. This step effectively increases the average length of the ESTs in the original master EST list. By constructing longer sequences, one increases the amount of information available when assessing the significance of a given hit. Each contig can then be compared to all known DNA and protein sequences. If the sequence with the best score is a member of the original protein family, then the contig is likely to represent either a known family member or a homolog.

In addition to searching ESTs directly with the sequences of the proteins in the family of interest, one can retrieve the annotations available for each protein in the family. Such annotations can be parsed for database cross-references, literature references, descriptive key words, and characterizing motifs. This information can be incorporated into Overview in the form of properties.

Finally, one can define aggregate properties of a protein family by constructing a multiple sequence alignment. A multiple sequence alignment is useful for defining profiles and motifs. Profiles and motifs can be used to search an EST database to find interesting ESTs as described above.

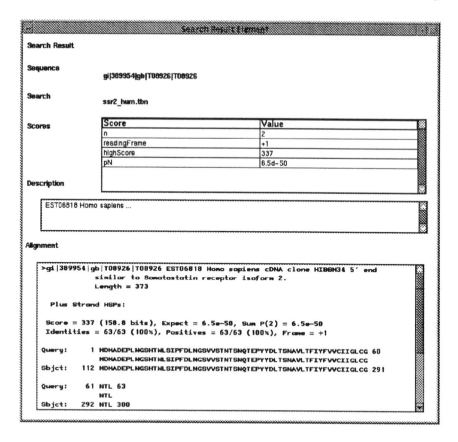

Figure 10.9 SearchResult element: EST match to a protein superfamily. To find ESTs coding for proteins in a superfamily, use one member of the family, for example, human somatostatin receptor (SSR) type 5, as a BLASTP query. Use the top matches from this search as queries for TBLASTN and TFasta searches of EST databases. Parse the results of these searches into an Overview cluster (see figure 10.3). Take intersections of these searches (see figure 10.8) or sort the combined SearchResult elements (see figure 10.4) to show which EST sequences match more than one member of the superfamily. EST06818 (shown here) is the best match; it had already been identified as similar to SSR2. No other ESTs had strong matches.

10.5.2   Characterizing a Set of Novel Sequences

Given an uncharacterized sequence, a researcher might want to find similar sequences to make some guesses about its biological function. These guesses can be used as hypotheses for bench experiments (i.e., to predict how a system may change if the gene is altered). The user can specify a subject sequence (i.e., probe) by referring to all or part of an existing sequence derived from a sequence database, a laboratory information management system, or a file or by manually typing the sequence into the system.

The uncharacterized probe sequence can then be compared to a set of known, characterized sequences. The SearchList against which the probe is compared can be defined as all sequences in a given repository (e.g., GenBank or Swiss-Prot) or pointers to individual sequences in one or more repositories (e.g., a list of all human sequences of known function generated by filtering one or more SequenceDBs).

The comparisons can be defined as a series of similarity tests, using several different algorithms and several parameterizations of each, from fast heuristic methods such as BLAST to slower, more rigorous methods such as Smith-Waterman. It is desirable that the series of comparisons can be monitored and stopped once sufficient progress has been made. If a researcher wants to examine comparisons that have already been run, it is desirable to browse through the previously generated results rather than redo the comparisons.

Once the set of comparisons have been completed, one can analyze the results. Analysis involves looking for (1) close matches predicted by any one algorithm and parameterization, (2) distant matches that occur in more than one method, and (3) close matches or consistent weak matches to a fragment of the probe rather than the whole probe. The final step is to form a hypothesis for experimental benchwork from close or consistent matches to sequences of known function.

10.5.3   Analysis of a Genomic DNA Fragment

Positional cloning is a process that involves mapping disease genes to a specific region on a chromosome using linkage analysis. To map a gene, a population of related, affected people is required. If a genetic marker that correlates with the disease can be found in this population, then it may be possible to focus a search for a relevant gene to a relatively small region of a chromosome (1–2 megabases). To date, nearly 100 genes have been identified using this strategy.

If the disease-linked region has been sequenced, one can apply a gene-finding algorithm (Xu et al., 1994) to identify exons (i.e., regions of DNA that code for protein). These exons can be used in a functional characterization strategy as described in section 10.5.2.

10.6   Future Work and Generalizations

Specific areas for further development of Overview include more directly invoking programs and accessing databases, using database technology rather than flat files for persistent storage and concurrency, more system- and user-defined filters for search result elements, and including an intuitive comparative measure of sequence comparison ratings across different programs. We also want to (1) validate the functional requirements and user interface by preparing test data for a simple but thorough case study, (2) install those data in the laboratory, and (3) observe researchers as they use the system. The current prototype is installed at the University of Washington Biotechnology Lab, Zymogenetics, and the T4 lab at Evergreen. An open question is the extent to which researchers will trust software to track their results. Experience to date at the Environmental and Molecular Research Center (Pacific Northwest National Laboratory) indicates that scientists using beta versions of that software seem willing to trust the database to hold experiment results and do not independently track particular input and output files. While we cannot report on how successful ECCE will be in the field until the alpha versions are released next year, initial user acceptance is high.

Overview aims to be an intelligent workbench environment that will enable researchers to apply multiple comparison and analytical algorithms to a set of data and provide an interface for the analysis and evaluation of the various results within this broader context. Given a new DNA or protein sequence, a scientist will typically compare the sequence to a repository of known sequences (e.g., Swiss-Prot or GenBank) using one of a number of comparison algorithms (e.g., Fasta, BLAST, or Smith-Waterman). Each comparison algorithm has a large number of parameters that a user can change. Unfortunately, there have been few quantitative analyses that clearly demonstrate how one method and parameterization differs from another when applied to real-world data. A first step has been the development of a statistical test using an extensive query set derived from the Protein Information Resource (PIR); this test can be used to evaluate the relative performance of multiple comparison methods and parameterizations (Shpaer et al., 1996). More important, however, we aim to understand how the various methods complement one another when used together.

We believe that Overview will not only help track results but also provide insights for specialized functions and heuristics to reorder sequence comparison results according to problem-specific criteria. Our ability to generate and store the results of various sequence comparison programs has been the focus of our initial efforts. We look now toward developing ways to provide meaning to the sorting and meta-analysis of these data— knowledge discovery is our goal, not data archiving. Therefore, Overview must provide us with an avenue to understand what it means when different algorithms attempting to answer the same question (are there significant similarities between sequences?) provide significantly different rank orders of results (i.e., the sorted order of "scores" of the two algorithms provides a different view of which comparisons are "significant"). This question is critical since the ability to discover distant homologs, as indicated by lower similarity scores, often provides the most insight as to the functional role of the sequences as well as experimental systems in which to test these hypotheses. For example, discovering unambiguous similarity between two human sequences may be interesting, but if nothing is known of the function of either one, then the investigation has nowhere to go. On the other hand, if a new human sequence is found to be similar to a sequence from a yeast or bacteria for which we do have some functional understanding, we can immediately ask if our new human sequence has an analogous function. Since yeast and bacteria are significantly easier to experiment on, we are also provided a convenient platform from which to test our new model. These questions are critical to both basic biological understanding as well as the development of novel tools and therapeutic systems.

Further work toward solving the domain-specific research challenges above are facilitated by Overview's ability to track the results of sequence comparisons and to organize these results flexibly. Overview, by providing infrastructure for the organization of results, should assist molecular scientists in developing algorithms for organizing, categorizing, and classifying search results across several different applications. Technological barriers to this work remain, however. Computer science research that could be brought to bear or extended to address these barriers include

1. data mining algorithms to find similarities in search result elements;
2. articulation of an appropriate software architecture for attaching new sequence search applications, which are essentially domain-specific queries on a large heterogeneous database;
3. the use of semistructured techniques (Buneman et al., 1995; Abiteboul et al., 1997) to parse the output of computational applications and provide better extensibility to our system;
4. specification of user-defined functions that can be dynamically added to the system to extend and specialize built-in comparison algorithms, sorts, and filters; and
5. development of a scripting language to automate repetitive tasks— ideally, this language would permit the capture of a sequence of user operations to "replay" (and possibly edit) operations that have been run on another set of search result elements.

The current Overview implements, in effect, a user-oriented, high-level and flexible implementation of lists. It provides researchers a way to organize results into clusters and to view lines of results from different programs as separate entities in a consistent format. Our requirements analysis has shown this to be the most immediately useful results-tracking or data-mining activity that we could provide. No existing system that we could find provides these capabilities, despite the ubiquitous nature of lists and sets as common data structures for programming.

While some issues of interoperability, such as incorporating new information in the systems data model and adding new programs or databases without new programming effort, are still significant research areas, increased interoperability will increase the need for tools to track and view many more computational results than researchers now deal with. Overview is an effort to explore new requirements introduced by advances in interoperability.

Acknowledgment

A shorter version of this chapter appeared as Cushing et al. (1997). We are thankful to the conference attendees and the program committee of the Ninth International Conference on Scientific and Statistical Database Management (Olympia, WA, 1997) for many helpful suggestions.

Chapter 11

RNA Structure Analysis: A Multifaceted Approach

Bruce A. Shapiro, Wojciech Kasprzak,
Jin Chu Wu, and Kathleen Currey

Genomic information (nucleic acid and amino acid sequences) completely determines the characteristics of the nucleic acid and protein molecules that express a living organism's function. One of the greatest challenges in which computation is playing a role is the prediction of higher order structure from the one-dimensional sequence of genes. Rules for determining macromolecule folding have been continually evolving. Specifically in the case of RNA (ribonucleic acid) there are rules and computer algorithms/systems (see below) that partially predict and can help analyze the secondary and tertiary interactions of distant parts of the polymer chain. These successes are very important for determining the structural and functional characteristics of RNA in disease processes and in the cell life cycle.

It has been shown that molecules with the same function have the potential to fold into similar structures though they might differ in their primary sequences. This fact also illustrates the importance of secondary and tertiary structure in relation to function. Examples of such constancy in secondary structure exist in transfer RNAs (tRNAs), 5s RNAs, 16s RNAs, viroid RNAs, and portions of retroviruses such as HIV. The secondary and tertiary structure of tRNA Phe (Kim et al., 1974), of a hammerhead ribozyme (Pley et al., 1994), and of *Tetrahymena* (Cate et al., 1996a, 1996b) have been shown by their crystal structure. Currently little is known of tertiary interactions, but studies on tRNA indicate these are weaker than secondary structure interactions (Riesner and Romer, 1973; Crothers and Cole, 1978; Jaeger et al., 1989b). It is very difficult to crystallize and/or get nuclear magnetic resonance spectrum data for large RNA molecules. Therefore, a logical place to start in determining the 3D structure of RNA

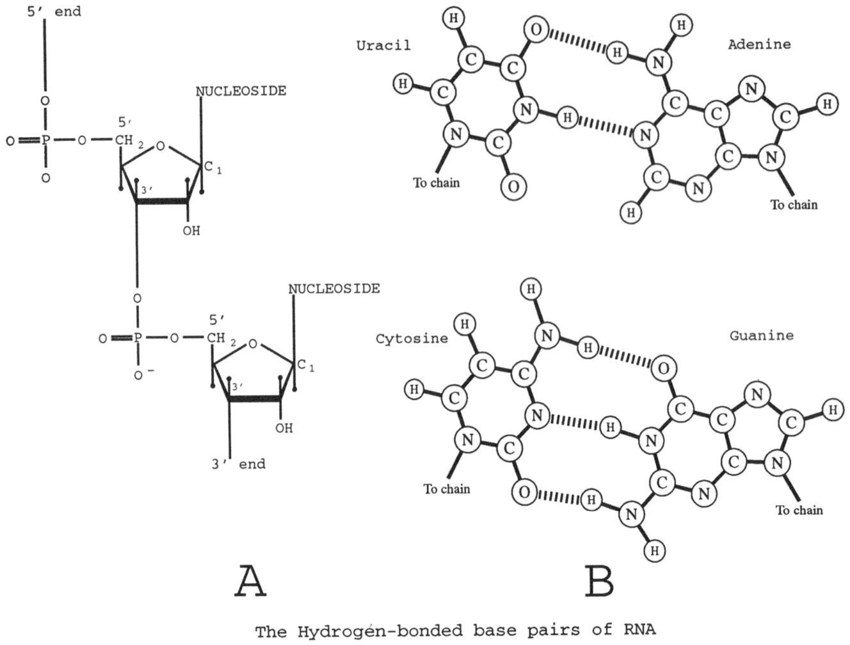

The Hydrogen-bonded base pairs of RNA

Figure 11.1 A schematic illustration of the RNA backbone (a) and nucleotides, shown as base pairs A–U and G–C (b).

is computer prediction of the secondary structure. The sequence (primary structure) of an RNA molecule is relatively easy to produce. Because experimental methods for determining RNA secondary and tertiary structure (when the primary sequence folds back on itself and forms base pairs) have not kept pace with the rapid discovery of RNA molecules and their function, use of and methods for computer prediction of secondary and tertiary structures have increasingly been developed. Accurate determination of RNA structure requires an understanding of basic interactions such as hydrogen bonding, stacking, and hydration in differing structural contexts.

11.1 A Brief Introduction to RNA

RNA is a molecule that can play several different roles in nature. Chemically it is very similar to DNA (deoxyribonucleic acid) but is different enough to make it structurally and functionally quite distinct from DNA. RNA molecules are polynucleotides containing ribose sugars connected by 3′–5′ phosphodiester linkages. The bases composed of A (adenine), C (cytosine), G (guanine), and U (uracil) are connected to the ribose sugars, which in turn are attached to the sugar-phosphate backbone, each forming a nucleotide (see figure 11.1). An RNA chain is most often single stranded as opposed to double stranded, as is the case for DNA. These single-stranded RNA sequences have a tendency to fold back on themselves to form double-stranded structures. Most commonly, Watson-Crick base pairs between A and U and between G and C are formed, which involve two and three hydrogen bonds, respectively. Less stable pairings are also possible, such as G with U. The consecutive formation of base pairs results in the formation of a stacked double helix. In addition, the unpaired regions produce morphologic features commonly called bulge loops, hairpin loops, internal loops, and junctions or multibranched loops (see figure 11.2). Structures formed in this way are called *RNA secondary structures*.

Also, *tertiary interactions* are possible whereby free base regions in loops or dangling ends may interact with one another forming pseudoknots (Pleij et al., 1985; Pleij, 1990; ten Dam et al., 1992) as well as more general interactions. Consequently, a pseudoknot is always characterized by at least two stems. If all four types of single-stranded loop regions are taken into account, 14 different types of pseudoknots can be theoretically predicted, although not all are sterically possible. Examples include base pairing between a hairpin loop and a single-stranded region, a hairpin loop and bulge loop, a multibranch loop and multibranch loop, and so forth. Almost all known pseudoknots involve stems of three base pairs or more, and many of the pseudoknots described to date are H-type or hairpin pseudoknots. These are formed when hairpin loops interact with single-stranded regions, not within other loops. Pseudoknots now appear to be a common structural motif in a number of RNAs, including viral RNAs (Pleij, 1990; Jacobson et al., 1993). Triple base interactions have also been described and occur when nucleotides in single-stranded positions interact with base pairs. Determining triple base interactions at this time is not feasible with computer prediction methods, although methodology using comparative sequence analysis has been reported to help identify potential base triples (Gautheret et al., 1995).

Ultimately, it is the combination of the secondary and tertiary structural interactions as well interactions with the environment, including water, ions, and proteins, that produces the 3D arrangement of the RNA and RNA-protein complexes. It is this 3D structure that determines the molecule's function. Some of these functions are outlined below (for further information, see Watson et al., 1987).

RNA viruses known to cause infections include HIV, rhinovirus, poliovirus, measles virus, and ebola virus, to name a few. In some cases the disease state is accomplished by a process known as *reverse transcription*. Here, the RNA is back-translated and incorporated into the cell's DNA. This new genetic information then alters the cell's machinery in various ways, allowing the virus to grow in the body and in some cases to cause cell death. The formation of viral particles as well as various viral control mechanisms is related to the structural properties of the molecule.

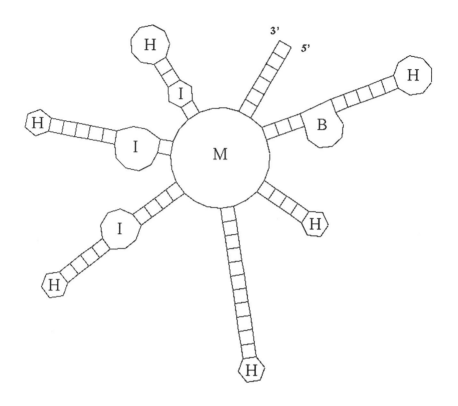

Figure 11.2 Morphologic features of RNA secondary structures: hairpin loops (H), bulge loops (B), internal loops (I), and junctions or multibranched loops (M).

RNA is also a major component of the cellular machinery for protein production. Messenger RNA (mRNA) is produced from the DNA. The mRNA in turn serves as a template by means of triplets of nucleotides (*codons*) to code for the individual amino acids, which in turn produce the polypeptide chains that form the proteins. The rates of expression of these proteins are controlled by the structural characteristics of the mRNA.

Transfer RNA (tRNA) is another form of RNA that is very important in the cell life cycle. This molecule also participates in the process of protein production by providing an anticodon for specific amino acids that are used to build a protein chain. There is a different tRNA for each type of amino acid. Thus, there is a correspondence between the codons in the mRNA and the tRNA. Again, the structural characteristics of the tRNA control its function. Recent data also suggest that disease states may exist due to mutations in mitochondrial tRNA.

Ribosomal RNA (rRNA) is yet another RNA molecule that participates in the protein production process. This molecule is a part of a nucleoprotein complex that consists of both proteins and rRNA and is called a *ribosome*. It is on the ribosome that proteins are synthesized. In this case, the molecule is acting as a structural molecule as opposed to a carrier of genetic information. Eighty-five percent of cellular RNA is made up of rRNA.

Recently, discoveries have shown that RNA can act as an enzyme or a catalyst—a function that was previously relegated to proteins. This enzymatic activity occurs as a function of structural as well as sequence-dependent characteristics of the molecule. This is significant for RNA splicing; moreover, this important property has sparked research into genetically engineered RNA drugs to cut RNA viruses at specific locations, thus destroying their ability to reproduce.

11.2   RNA Folding Problem

RNA secondary structure prediction is computationally quite intense. It is estimated that there are 1.8^n possible structures for a sequence of size n. The enormity of the task is immediately apparent when one considers an RNA sequence such as HIV, which is over 9,000 nucleotides in length. The goal of RNA structure prediction is to locate the biologically relevant structure(s) in this huge landscape.

RNA foldings are usually computed to minimize free energy. At a given temperature, the molecule is assumed to be in equilibrium, and a minimum energy solution can be regarded as a probable folding. However, one has to take into account folding pathways and the molecule environment, which also have a significant impact on the resultant structure. Therefore, the functional state may not necessarily be the minimum energy state. Energy is assigned to hydrogen bonds between bases and to the stacking of one base over another. Various loops are given destabilizing energies. A principle of additivity is assumed, so that the overall free energy of a folding is the sum of the energies of the stacked base pairs and the loops. This assumption is essential since better experimental data are not yet available to provide more suitable rules. Two types of algorithms are used within Structurelab (Shapiro and Kasprzak, 1996) to determine the folding of the molecules. These are a *dynamic programming algorithm* (DPA) and a *genetic algorithm* (GA), both briefly described below. Both of these algorithms run on a MasPar MP-2 supercomputer; however, Mfold (the dynamic programming algorithm) is also available on a variety of computer architectures.

The time requirement for predicting RNA secondary structure by the commonly used Mfold is of order $O(n^3)$, where n is the length of the sequence. The availability of a machine such as a MasPar computer with a massively parallel architecture changes the time relation to length from $O(n^3)$ to $O(n^2)$. For example, a sequence of size 2,000 bases takes about 1,583 s to fold on a single-processor CRAY-YMP, while the same sequence takes 1,149 s on a MasPar MP-2. Increasing the size of the sequence to the length of a strain of HIV (9,128 bases) causes the fold to take approximately 103,344 s on a CRAY-YMP and 23,201 s on a MasPar MP-2 (Shapiro et al., 1995). Computer time requirements for tertiary structure prediction are even more demanding and therefore push the limits of current technology. Deterministic algorithms such as Mfold are not able to predict tertiary structure; rather, they are solved in nondeterministic ways such as with the genetic algorithm implemented on the MasPar (see below).

Recursive programs, such as Mfold, build optimal foldings one base at a time. They consist of two distinct parts. The first part, called the "fill," computes and stores minimum folding energies for all fragments based on minimum folding energies of small fragments, starting with pentanucleotides. The final part, called the "traceback," assembles a structure by searching through the matrix of folding energies. The fill algorithm takes the bulk of computing time, while the time for a single traceback is relatively negligible. However, to reasonably explore the conformational space of the foldings, several thousand tracebacks may be needed. The number of tracebacks depends on the size of the given sequence and the energy range one wishes to explore (Jaeger et al., 1989b). Computer memory requirements, which grow as the square of the sequence length, usually limit this kind of algorithm. Pseudoknots cannot be handled, and the additivity assumption for free-energy contributions is necessary. By their nature, recursive algorithms predict the optimal folding of all subfragments. This means, for example, that optimal foldings of a growing RNA sequence can be simulated without any additional cost, once the entire sequence is folded. Recursive algorithms are designed to yield a single solution, but with some modification the algorithm can be extended to yield multiple solutions (Zuker and Stiegler, 1981; Zuker, 1989, 1996).

A recently developed algorithm for secondary structure prediction is a genetic algorithm (GA). It is a nondeterministic algorithm that is implemented on a massively parallel MasPar MP-2 supercomputer. The MasPar is an SIMD (single instruction multiple data) machine with 16,384 processors. The algorithm is also currently being ported to a Cray T3E MIMD (multiple instructions multiple data) machine. The GA iterates, in parallel, using a three-step evolutionlike procedure: selection, mutation, and crossover on all 16K processors. When the genetic algorithm is started, an initial stem "pool" (thousands of possible base pairing regions) is created, and then the selection operator chooses a set of best fit structures from the current population. From this selected set a new generation is created. Mutations induce random changes in the RNA structure at each generation. Crossover allows structural information within the population of structures to be exchanged, creating new structures based on the mutations and exchanged structural characteristics. Each generation iterates through this process of selection, mutation, and crossover. Eventually the population converges to structures that are highly fit. The fitness criterion used is free energy. Pseudoknots are allowed, and the current version of the algorithm is capable of predicting H-type pseudoknots (Shapiro and Navetta, 1994; Shapiro and Wu, 1996, 1997).

Classically, phylogenetic analysis has also been used as a method for secondary structure determination. Natural phylogenetic analysis assumes that RNAs of similar function share common structural elements. These elements have been shown potentially to exist by compensatory mutation of bases in a double helix, for example, a G–C to an A–U pair. Sequence phylogeny cannot predict standard base pairs well, since the rules for these pairs are not precise. A phylogenetic analysis compares sequences and looks for similar secondary structures. Alignments are helpful for studying sequence homology and usually contain gaps to compensate for base deletions and insertions. Secondary structures determined from the alignment must show at least one compensatory base pair mutation per helix in order to be confirmed by this method. Tertiary interactions such as pseudoknots and triple base interactions have been proposed once the secondary structural elements were established. Tertiary base pairings have also been shown to exist from compensatory changes, just as in case of secondary structure.

Even with supercomputers (massively parallel machines and Cray-like machines), the only tertiary structure elements that can be predicted to date are pseudoknots. Two programs predicting tertiary structures are discussed in this chapter. One has the ability to identify all possible tertiary interactions between unpaired regions in a secondary structure. The other, the GA, in addition to predicting RNA secondary structure can predict H-type pseudoknots (Shapiro and Wu, 1997). The GA for pseudoknots calculates the free energy of the individual pseudoknots in addition to the free energy for the entire structure containing the pseudoknot(s). Another program predicts all possible pseudoknot base pairings (Martinez, 1990). It does not eliminate overlapping structures and does not calculate free energies of the structures.

Confidence in structural predictions is enhanced when used in conjunction with biological and phylogenetic data. The existence of electron microscopy, X-ray crystallography, and nuclear magnetic resonance as well as biochemical techniques have all contributed to elucidate the structure of RNA molecules. The remainder of this chapter deals with the computer methodologies that are used in Structurelab, a computer workbench that may be used with external experimental biological information to determine the structural attributes of RNA and to make these attributes more interpretable by the individual researcher.

11.3   The Genetic Algorithm: An Overview

As it already stated above, a new folding algorithm has been developed, based on the concepts of genetic algorithms (GA; Holland, 1975; Davis, 1991), and implemented on our massively parallel MasPar supercomputer. Our GA uses the capabilities of a parallel SIMD architecture, where local interactions occur in parallel and propagate throughout the population, using parallel mesh communication, thus causing completely different populations to interact. The algorithm iterates over a three-step, evolutionlike procedure: selection, mutation, and crossover, using minimal free energy as a criterion to improve structures across all processors at each generation. An iteration constitutes one generation. During a preprocessing phase, the GA generates a stem pool consisting of all possible fully zipped (or partially zipped) stems from a given sequence. After that, very simple structures are initialized across all processors. At each generation, in each processor, the GA selects two RNA structures from the processor and its eight neighbors (assuming the eight-way interconnected mesh architecture) using a ranked rule (biased toward the better free energies; Goldberg and Deb, 1991) and takes them as parents P1 and P2. Then the GA mutates the RNA structures by randomly picking stems from the stem pool, according to an annealing mutation operator (Shapiro and Wu, 1996), to form two child-structures, C1 and C2, excluding conflicting structures but identifying and including (if needed) stems that form pseudoknot interactions (see below). Next the GA does a crossover operation between (P1, P2) and (C1, C2) by distributing stems from P1 and P2 to C1 and C2 to complete the two new structures. Between these two structures, the one with the better energy is selected. Thus, 16,384 new structures are created in parallel at each generation. This process is repeated for each generation. Eventually the algorithm will terminate when the population becomes stable. The GA is typically run several times, and a consensus of structures is then used for further analysis (Shapiro and Wu, 1996).

In our GA, an annealing mutation operator was designed to deal with long RNA sequences that create large stem pools. It can also be applied to short RNA sequences. The mutation probability, p (i.e., the number of mutations per stem per processor), descends along a hyperbola with respect to the size of the structure, s (i.e., the number of stems comprising the structure). That is, $p = (a/s) + b$, where a and b are empirically determined constants and b is less than zero. Hence, the total number of mutations across all processors at each generation drops linearly. Such an annealing mutation operator allows a fairly large number of mutations to take place across all processors when the process starts and reduces the total number of mutations at each generation as the sizes of the structures increase. In other words, it cools down the process slowly as the procedure continues (Metropolis et al., 1953; Press et al., 1992). Moreover, the number of mutations per processor at the very beginning of the process is assumed to depend on the size of the stem pool generated from the sequence. For larger stem pools created by longer RNA sequences, more mutations are allowed to occur at the beginning of the process. Thus, especially for long sequences with thousands of nucleotides, as opposed to hundreds of nucleotides, the annealing mutation operator can cause the distribution of free energies over all processors on a MasPar MP-2 to converge after only hundreds of generations in contrast with an undetermined amount of generations. In other

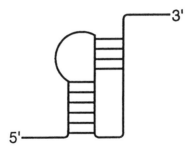

Figure 11.3   Example of a simple H-type pseudoknot.

words, eventually most processors out of a total of 16,384 processors contain the same structures. Consequently, based on this annealing mutation operator, a technique to terminate the GA was also developed using statistical methods (Shapiro and Wu, 1996).

The GA, besides its ability to predict secondary structure, is also able to predict H-type pseudoknots (Shapiro and Wu, 1997). These pseudoknots are formed by a hairpin loop pairing with a free strand to form a new stem. This new stem is adjacent to and coaxially stacks with the stem of the hairpin loop to form a quasi-continuous double helix (see figure 11.3). Such coaxial stacking contributes to the stability of the tertiary structure. Moreover, the two connecting loops consist only of free nucleotides. The connecting loop following the stem, which is closer to the 5' end of the sequence, crosses the *deep groove* of the RNA helix, and the other one crosses the *shallow groove*.

When dealing with pseudoknots, two issues are encountered: the topological structures of the pseudoknots and the related energy rules. Two stems that are not conflicting, that is, the sides of which are not overlapping, can form a pseudoknot if and only if one side of a stem is outside the bound of the 5' end and the 3' end of the other stem while the other side of the stem is within the bound. Such a relation is reciprocal for both stems. Given a tertiary structure of an RNA sequence, without any a priori knowledge, it is very hard in most cases to determine the order of stem formation that comprises a pseudoknot. This is because of the relative positions of stems and loops in the tertiary structure. Such tertiary interactions are governed not only by geometric factors, such as length requirements of the connecting loops crossing the deep and shallow grooves, but also by chemical and thermodynamic factors. Thus, the topological structure formed by base pair interactions would have to be determined.

Pseudoknots are generated in the algorithm by keeping two lists of stems on each processor in the order of their 5' ends. One list contains stems constituting a secondary structure of an RNA sequence, and the other list stores pseudoknot stems that complete the formation of pseudoknots. At each generation of the GA, after forming the intermediate secondary structures out of the first list of stems, the possible H-type pseudoknot stems are added into the structures. Thus, at each generation, structures that contain pseudoknots compete with those that do not. It is a deterministic process to produce a secondary structure from the first list of stems using a tree representation (Shapiro, 1988; Shapiro and Zhang, 1990). With this tree representation, the sizes and relative positions of free strands, stems, and loops in the secondary structure can be precisely determined. When added to a secondary structure, the relative geometry of the pseudoknot stems with respect to all morphological components in the secondary structure can be thereafter measured. These structures can be visualized and analyzed in 3D using rna_2d3d (see below).

Once the topological structures of the H-type pseudoknots are resolved, the free energy of a given tertiary structure, acting as the fitness criterion of the GA, is determined by the related energy rules. An appropriate energy rule can eliminate some structures that do not satisfy stereochemical or thermodynamic restrictions. This can be accomplished by the selection procedure and the competition between the two child structures after the crossover procedure at every generation of the GA.

Three different energy rules for secondary structures (Freier et al., 1986; Turner et al., 1988; Woese et al., 1990) are employed in our applications. There is still a need for energy rules that can be applied to the variety of pseudoknots. More experiments need to be performed to determine these rules. For the H-type pseudoknot, a simple energy rule (Abrahams et al., 1990) is used. The stable H-type pseudoknot can have only connecting loops shorter than 16 nucleotides and 4.2 kcal/mol is assigned to each loop, which destabilizes the structure of an RNA sequence. Thus, the free energy of a pseudoknot is the sum of the energies of the two stems involved, their stacking energy (which includes the hydrogen bond energy), and the energies of the two connecting loops. Other genetic algorithms have also been developed (Gultyaev et al., 1995; van Batenburg et al., 1995).

11.4   Structurelab: An RNA Structure Analysis Workbench

An RNA structure analysis system called Structurelab (Shapiro and Kasprzak, 1996) has been developed and is being continually expanded to include more functionality, which allows a researcher to interactively and methodically pursue a multiperspective analysis of RNA structural conformations. The process has involved the development of new algorithms to explore secondary structure motifs and RNA tertiary structure and provided methods for the measurement and visualization of structural similarity from the global and contextual points of view. The system includes facilities allowing exploration in detail of both multiple and individual RNA structures. This approach views the structure determination problem as one of dealing with a database of many computationally generated structures and provides the capability to analyze this dataset from different perspectives. Consequently, databases containing thousands of structures can be analyzed. Functions also exist to permit the analysis of RNA tertiary (3D) substructure interactions.

The system utilizes various software modules and hardware complexes available both locally, at the Frederick Cancer Research and Development Center (Frederick, Md.), and elsewhere through computer networking. These include Cray Y-MP, Convex, MasPar SIMD massively parallel machine, VAX, and a number of SUN, SGI, and DEC workstations. Employing different computational platforms makes it relatively easy to incorporate already existing programs and to add newly developed algorithms and best match these algorithms to the appropriate hardware. Since the field of bioinformatics has been rapidly expanding, it is only reasonable to expect the appearance of new software tools running on any or all of the available platforms of our local area network or on some distant machines. Therefore, part of the idea of our ongoing project was to develop a computer workbench providing a uniform and user-friendly interface to the abovementioned packages as well as new tools for visualization, manipulation, and analysis of the available data and data-processing results.

11.4.1   Implementation

The system grew out of work done in the 1980s (Shapiro, 1988; Margalit et al., 1989; Shapiro and Zhang, 1990). A flexible, robust, and portable communications package linking a user-friendly interface with the remote and integrated analysis packages is provided as a central part of the system. It encapsulates the knowledge about the characteristics of the network. It is modifiable as well as expandable to accommodate changes in the "environment." It also attempts to make the underlying network complexity, architecture, and operating system variety as transparent to the user as possible. For interactions among distributed portions of the system, the communications package utilizes a form of message passing.

Functions invoked on a central controller workstation can be run locally or on other platforms of the network. In some cases only one platform supports a given function/application, but in the majority of cases the user can select which of the available platforms should be employed. The results are reported back to the central controller, where they are interpreted and displayed on the user's front-end machine. This is schematically illustrated in figure 11.4.

To facilitate additions of new architecturally dependent software, we have made our system "portable," by which we mean able to expand and/or move the RNA Analysis Workbench from one physical network onto another

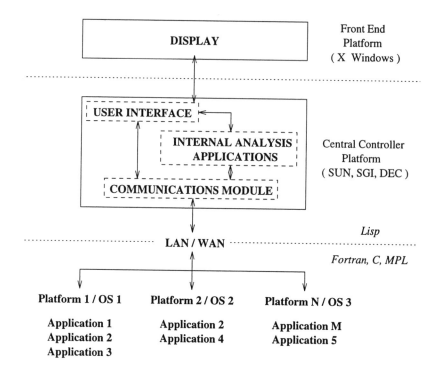

Figure 11.4   Structurelab: layout of a system utilizing a network of heterogeneous computing platforms.

with comparable types of machines and operating systems, and not necessarily able to run it on any machine on a network. We have successfully
demonstrated the system's portability by installing it on a temporary network set up for the Supercomputing '94 exhibit in Washington, D.C., utilizing MasPar's floor machine as well as our own MP-2 in Frederick, Maryland
(50 miles apart). Names and IP addresses of the machines employed were
defined on the show floor.

A hierarchy of menus constitutes a core of the graphical user interface
to the available functions. The user can select any source and target machines within a network defined in reconfigurable tables, and the system
will handle setting up the required connections, file transfers, batch job
submissions, or real time interactions. All the menu items have terse on-
line help information associated with them that explains the basic concepts
and gives usage hints. Figure 11.5 shows a selection of the system tools as
they appear to the user.

One more important feature of the system is the distributed error trap-
ping and handling. It attempts to allow the user as many run-time correc-
tions of errors as possible. At the very least, after aborting execution, it
returns to the last consistent state of the system.

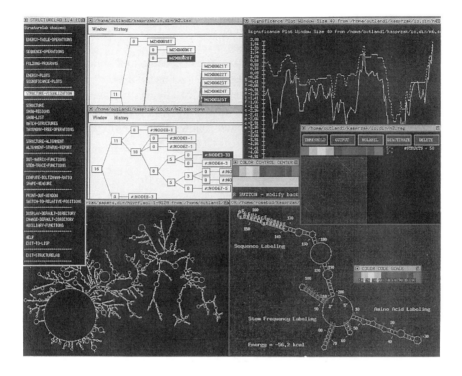

Figure 11.5 Structurelab: a general view of some of the available workbench tools. Shown are (clockwise, starting from upper left) the main menu, the taxonomy tree windows, the significance/stability plot, the 2D stem histogram, the structure drawing with base labeling, amino acid labeling and annotations, the interactive color control window and a small color-scale window, and the large-scale structure drawing.

11.4.2 Specific Application Characteristics

The following list of specific applications does not exhaust all functions available in Structurelab. However, the major functional domains are presented.

Sequence Manipulation

A set of interactive applications running on the central controller workstation allows the user to manipulate and design sequences with special characteristics. These may be folded and the resulting secondary structures compared with those predicted for the wild-type sequences. In this way sequence and structure dependence can be studied in a variety of ways.

Functions available in this group perform sequence extraction from a database (such as GenBank), format conversion, and customized sequence creation. They also facilitate sequence manipulation via splicing operations and several types of mutations that permit specific and random mutations of single bases or specified intervals, with options to preserve amino acids or exclude selected codons from mutations. Finally, translation to amino acid sequences (protein building blocks) is available. A typical sequence format, among several possible, is illustrated below (line format, name, sequence length, sequence).

```
(60A1)   sequence-name.zuk
115
GAAUUACCGAUAUCGAUACAUCAGGAAUAUUUGAUUCAGAUGAUAUGACUAUCAAGGCCG
CCUGAGUGCGGUUUUACCGCAUACCAAUAACGCUUCACUCGAGGCGUUUUGUGCG
```

Folding Algorithms

Structurelab currently employs two different types of folding algorithms: the dynamic programming algorithm (DPA; Zuker and Stiegler, 1981; Turner et al., 1988; Jaeger et al., 1989a, 1989b; Zuker, 1989; Shapiro et al., 1995) and the genetic algorithm (GA; Shapiro and Navetta, 1994; Shapiro and Wu, 1996, 1997). While both attempt to predict the best secondary structures for a given RNA sequence, using free energy rules as criteria, they differ in basic concepts, as explained above. In addition, the GA offers a pseudoknot prediction capability. Other RNA folding algorithms may in time be included (Mironov et al., 1985; Martinez, 1988).

The RNA folding algorithms accept sequence files (strings) as input, and output multiple region tables indicating which bases (nucleotides) pair with which ones in a folded structure (see table 11.1). These region tables reflect energetically optimal and suboptimal solutions based on the standardized free energy rules. Figure 11.6 illustrates the fold of the entire HIV-1 sequence (strain HIVRF) generated by the DPA folding program on a 16,384 processor MasPar MP-2 computer in 6 hours 20 min.

Table 11.1   An example of a region table representing an RNA structure.

Region number	Start (first 5' base)	Stop (last 3' base)	Region size (base pairs)	Energy (kcal/mol)
1	2	425	9	−15.4
2	12	413	2	−1.7
3	19	79	3	−5.8
...	...	...	...	...

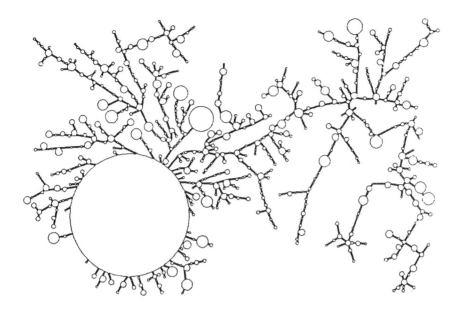

Figure 11.6 RNA secondary structure of an HIV-1 virus (HIVRF strain, 9,128 nucleotides long). Drawing was performed by our own visualization functions described in the text.

Structure Representation

Structural representation for a set of RNA molecules can be created based on the region files produced by the folding programs. Secondary (2D) structures are represented as trees, utilizing Lisp's nested list notation, with symbols such as H (hairpin loop), B (bulge loop), I (internal loop), and M (multibranch loop) along with property lists. Optional Rs present in some representations are no-op placeholders indicating regions (hence R) or stems (Shapiro and Zhang, 1990). These representations are very well suited to pattern matching and motif determination.

Some of the tree representations used are shown below.

1. Condensed format, loops only:

$$(N(H)(H)(BH)(H)(H)(H)(BBBIH)))$$

2. Expanded format, loops only:

$$(N(H)(H)(B(H))(H)(H)(H)(B(B(B(I(H)))))))$$

3. Expanded format, loops, and regions:

```
(N(R(H))(R(H))(R(B(R(H))))(R(H))(R(H))(R(H))
             (R(B(R(B(R(B(R(B(R(I(R(H)))))))))))))
```

4. Expanded format, region size, and loop component size:

```
Tree Representation
(N(R(H))(R(H))(R(B(R(H))))(R(H))(R(H))(R(H))
             (R(B(R(B(R(B(R(B(R(I(R(H)))))))))))))
Nsize 2;
rsize 8;
cHsize 7;
rsize 6;
cHsize 5;
rsize 3;
  ...
cIsize 2 1;
rsize 4;
cHsize 5;
```

Structure Alignment

Aimed at detecting structural motifs, this pairwise "Needleman-Wunsch" alignment function permits clustering of RNA secondary structures based on similarity of substructures (Needleman and Wunsch, 1970; Sobel and Martinez, 1986). It uses the parenthesized string form for representing trees, described above, and performs a multiple alignment clustering of such representations (Shapiro, 1988; Margalit et al., 1989). Shown at the top of the next page is an example of an alignment output file for suboptimal structures of a sequence called m2.

Taxonomy Operations

Taxonomy tree applications form a class of functions permitting calculation, graphical display, search, and manipulation of structural taxonomy trees. The fundamental application performs a fast pairwise tree comparison on the tree representations of the RNA secondary structures and generates a taxonomy tree that clusters the structures based on heuristic measures of similarity. The tree representations facilitate multiple levels of abstraction of the actual structure allowing for structural comparisons of varying strictness (Shapiro and Zhang, 1990). This includes tree shape comparison independent of loop or region sizes, tree shape using loop sizes, and tree

```
       Clustered Pairwise 'Needleman-Wunsch' Alignment
       in 'identity (no translation)' alphabet of

             1.  M2X00011T  (1-28)
             2.  M2X00017T  (1-28)
             3.  M2X00007T  (1-28)
             4.  M2X00012T  (1-33)
             5.  M2X00013T  (1-33)
             ...
       listed in clustered order.

   1 (N(I     IIbBIH)(H)( H)(H)(BBBIH))
     ||||      || ||||||||| |||||||||||||
   1 (N(I     IIIBIH)(H)( H)(H)(BBBIH))
     ||||      | ||||||||| |||||||||||||
   1 (N(I     IbIBIH)(H)( H)(H)(BBBIH))
     ||||         |||||| |||||||||||||
   1 (N(IH)(H)(H)(BH)(H)( H)(H)(BBBIH))
     ||||||||||||||||||||| |||||||||||||
   1 (N(IH)(H)(H)(BH)(H)( H)(H)(BBBIH))
     ||| ||||||||||||||||| |||||||||||||
                   . . .
     (N(ih)(h)(h)(bH)(H)( H)(h)(bbbiH))

   ALIGNMENT SCORE = 1342.00
```

shape including the sizes of components comprising the loop. Another function draws the tree and allows the user to search it, display the associated structure drawings, and compress and decompress the tree leaf nodes selectively (see figure 11.5).

The function may be used in conjunction with or in place of the above-described method based on string alignments. The tree representation allows for a more finely tuned control of the editing operation for comparisons. However, the Needleman-Wunsch technique gives an alternative and explicit depiction for motif discovery.

Structure Matching

The structure matching functions deal with motif analysis of a database of structures (possibly thousands). They provide the information on searched motif frequencies, which is especially helpful in the analysis of the large suboptimal solution spaces usually produced by the DPAs. It is also the most reliable tool at picking some rare (low-frequency) motifs that may be hard to find using other visualization-oriented tools.

Functions available in this class can be divided into two subclasses: one dealing with tree representation matching and the other with linear feature matching. The structure matching operates on the tree list representations and performs pattern searches for structural motif queries that may include wildcards: "?" for a single position "don't care" and "+" for multiple positions. The linear pattern-matching functions utilize data structures used in the RNA structure drawings, and they allow the user to perform base-by-base searches on conjunctions of linear patterns, including

- specific (pair at position) or general base pairing patterns—(1 0 1 +),
- bases that take part in these pairings—(G C C +),
- structural patterns—(R B R +), and
- energy patterns.

RNA Structure Visualization

This rather extensive set of functions facilitates the drawing of RNA structures as well as manipulation of the drawings for optimum presentation via rotation, resizing, bending/untangling, labeling, and annotating. The drawings are based on a sequence file and related region tables (Shapiro et al., 1982, 1984). In addition, drawings can be generated directly from stem traces or stem histograms (composite drawings), both described below.

A recent addition to this set of tools is a 3D visualization and analysis tool utilizing Hugo Martinez's rna_2d3d software package, which generates atomic coordinates from structures predicted by the various folding programs (H. Martinez, personal communication). Our system couples it with RasMol, which facilitates visualization of these 3D representations (Sayle, 1994; see figure 11.7).

2D Stem Histogram

This function produces a 2D histogram, color coded for frequency information, of all base pairs that exist in a database of optimal and suboptimal structures produced by one of the RNA folding programs. In other words, a 2D histogram gives a picture of how prevalent certain structural motifs are among large numbers of energetically optimal and suboptimal structures.

Combining the statistical base pairing matrix analysis with the energy-oriented significance/stability information helps provide a fuller picture of the reliability of molecule folding predictions (see figure 11.5).

As was mentioned above, structure drawings can be generated directly from stem histograms. Due to the cumulative nature of such histograms, these drawings represent composite structures combining most frequently encountered structural elements within the graphed space of solutions. The

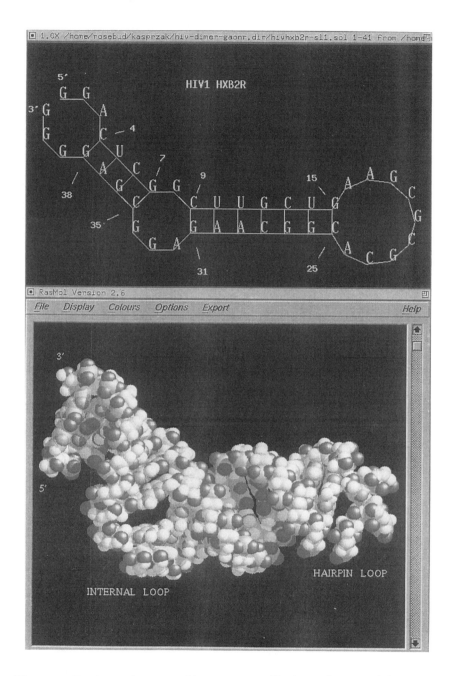

Figure 11.7 A sample stem and loop structure (SL1) visualization of the HIV-1 HXB2R strain dimerization site by our own secondary structure drawing function (top) and by RasMol, a 3D molecular visualization program (bottom). The secondary structure drawing is shown with base labeling and positional annotations, while RasMol depicts the atomic arrangements of the base pairs. The 3D picture shown is an unrefined depiction.

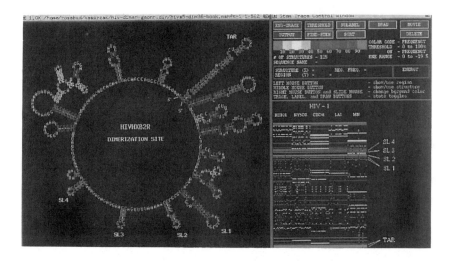

Figure 11.8   An example of a small multiple sequence stem trace depicting GA results for five strains of HIV-1 (25 runs each). A sample structure drawn and labeled interactively from the stem trace data is shown on the left. The single control window (upper right) provides the interface to potentially many stem trace windows. In normal use, sequence, graph positions, and energy and stem data are displayed in the control window. Labels within the stem trace window were added later for this illustration.

user can decide on the levels of frequency thresholding, with 50% being the lowest allowed in order to avoid stem conflicts in the same drawing.

Stem Trace

This function produces a 2D plot of all unique *regions* (*stems*) that exist in a database of structure (region) files. The generations (for GA traces) or suboptimal solutions (in case of DPA) are plotted on the horizontal axis, while all the unique regions are shown on the vertical axis. Persistence of structural elements thus can be viewed (see figure 11.8). The more continuous the horizontal line, the stronger the stem it represents.

Stem trace can be employed in a variety of ways. It permits a two-way analysis of GA structure predictions, depending on whether the user plots best solutions from consecutive generations of one GA run or the final generation solutions for multiple runs. In the first case the maturation of a structure predicted by the GA can be studied. In the second case variability of final solutions is illustrated. In the case of the DPA folding program results, stem trace facilitates visual exploration of the space of suboptimal solutions. Additionally, stem trace can be employed to compare folding results for multiple sequences on one graph. Once the sequences are aligned, the region tables produced by any of the folding algorithms can be appropriately adjusted and results plotted. In this way stem trace gives immediate visual clues about conservation of structural motives across a family of sequences. Thus, phylogenetic structural comparisons are possible.

Since a vertical slice through the stem trace graph is equivalent to a region table, the extracted data can be easily drawn for quick lookup of structural conformations or displayed as a movie sequence.

Tertiary Interaction Prediction

As mentioned in section 11.2, the GA can predict some tertiary structures (H-type pseudoknots). In addition, Structurelab provides a function that outputs lists of all potential loop–loop interactions, loop–free base interactions, and free base–free base interactions among the elements of the previously folded, 2D (secondary) RNA structure, used as input data (Le et al., 1991). Included in these are some validated pseudoknots (Pleij et al., 1985; ten Dam et al., 1992).

The user has a choice of several criteria by which the results are organized and printed out. A sample output is shown below.

```
DATE:  6/7/1994, TIME:  12:38:19

Sequence File : /home/outland1/kasprzak/io.dir/wtcar234.zuk
Region File : /home/outland1/kasprzak/io.dir/wtcar234.sol

SEARCH PATTERNS :

ENERGY Pattern:  ((>= 0.0) +)
HELIX Size Threshold:  3
INHIBIT Free-base/Free-base Interactions:  YES
OUTPUT Energy Threshold:  0.0
NUMBER of helices computed:  12
TOP N shown:  5
SORT key:  5' START

SEARCH RESULTS FOLLOW :

(58 119 3 -1.8)    5' Half Reg: H4 AAU    3' Half Reg: B1 AUU
(58 133 3 -1.8)    5' Half Reg: H4 AAU    3' Half Reg: H2 AUU
(71 132 3 -1.8)    5' Half Reg: M1 GUA    3' Half Reg: H2 UAU
(72 205 3 -2.6)    5' Half Reg: M1 UAC    3' Half Reg: I3 GUG
(74 134 3 -2.3)    5' Half Reg: M1 CAG    3' Half Reg: H2 UUG
```

Significance/Stability Plots

The significance function computes and graphs the stability/significance scores for secondary structure folds. Results are based on the energy of the optimal fold of a sequence window compared to the energies of a randomly shuffled and folded sequence within a window size chosen by the user (Le et al., 1988; Chen et al., 1990). Based on the positions of troughs in the plots, one can get a sense of the important structural elements.

This application has a batch job component (see below) and independent interactive graphing of selected data running on the central controller platform (see figure 11.5).

Energy Calculations and Energy Plots

This class includes functions that read data from energy table or tables (Freier et al., 1986; Turner et al., 1987; Woese et al., 1990) as well as energy-calculating functions. The data read in from the energy tables is then used by various other functions in the system. Both the newer Turner energy values in multiple-table format and older, single-table-based ones can be read in for the benefit of comparisons with older and newer structures.

The currently used energy table format (single or multiple tables) can be queried via a menu entry, and any rule sets can be read in to "reset" the energy data utilized by other functions. Energy calculations using any of the energy tables can be performed for the current structure or its interactively selected substructures. Another set of functions allows the user to plot and query for detailed output energy data extracted from the DPA folding algorithm's output file.

Miscellaneous Applications

A set of miscellaneous applications is provided to let the user perform "housekeeping," such as storing and restoring the system states, printing out display windows, and monitoring remote tasks. Various other functions, not necessarily strictly related to any specific application domain but available in many of them, belong to this category.

11.4.3   Classification of the Applications

The major utilities comprising Structurelab can be divided, based on the nature of communications associated with them, into batch and interactive classes. Some applications, such as the 3D drawing or the GA coupled with 2D, synchronized structure visualization, use both approaches in a single run. We refer to them as composite applications.

Interactive Applications

The interactive applications are characterized by user input and mouse interactions as well as synchronizing communications and file I/O capability across the network. Another characteristic of this class of applications is that exiting the chosen application domain ends data interaction.

This category includes

- sequence manipulations;
- generation of structure tree representations;
- pattern matching (pairing, linear, and subtree);
- drawing applications, such as 2D and 3D structure drawing, 2D stem histogram, stem trace, energy and significance/stability plots, and structure taxonomy tree;
- tertiary interactions, and
- miscellaneous user interface functions.

Batch Applications

Our use of the term "batch" application extends the usual definition and classifies as such not only jobs submitted to standard batch queues but also processes running without any need for synchronization after startup. Depending on the specific needs, the system automatically performs all input file transfers between source and target hosts prior to the actual remote process invocation or a job submission to a batch queue manager.

This category includes

- folding programs: DPA and GA,
- significance/stability calculations (for plotting),
- taxonomy calculations,
- structure alignment,
- energy data extraction (for plotting), and
- 3D coordinate calculations (which can be used in 3D drawings).

Composite Applications

A good example of a composite application in Structurelab is the 3D structure presentation. It starts with translating the output format of a folding program (DPA or GA) into a representation acceptable to rna_2d3d running on an SGI workstation. This program, invoked as a remote shell application but also waited on for synchronization purposes, calculates the 3D atomic coordinates of the folded molecule and outputs them into a Protein Data Bank PDB-formatted file. Once Structurelab detects the new PDB output file, it automatically performs the file transfer, if needed, to the front-end machine and invokes RasMol on it. This application facilitates interactive molecule visualization and manipulation.

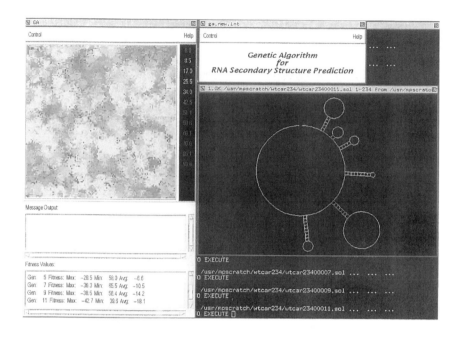

Figure 11.9 Genetic algorithm run under the Structurelab control. The left panel shows the fitness of structures in the MasPar's processor array, while the right panel illustrates the most fit structure present in the pool (or one randomly chosen from among several equally fit structures).

Another illustration of an application combining the two modes of operation discussed above and utilizing the flexibility of the system communications module is a remote GA run controlled from the central workstation and synchronized with the structure drawing application employed to show the progress of a GA run (see figure 11.9).

Interconnected Nature of Applications

Here we present a detailed discussion of interactive drawing applications to best illustrate the interconnected nature of the system tools. Many of the internal data structures are shared by the RNA structure drawing, 2D stem histogram, and stem trace functions. Therefore, it is logical that they can be employed to work in synchrony on the elements of the same RNA structure. Stem histogram and stem trace plots can be interactively queried for data to be overlaid on a structure drawing.

The RNA Structure Drawing can visualize multiple structures in a single run. Multiple sequence files and their related sets of region files can be read, based on a single input specification (using wildcards or explicit input lists), processed, and stored in a list of structures accessible via our internal directory. All the associated data, such as nucleotide sequence of the current drawing (i.e., the active window) and data structures needed to calculate the energy of the RNA structure or its elements, are part of the drawing state. It is updated with each selection of a drawing from the directory menu.

The 2D Stem Histogram visualizes structural motifs, base pair by base pair, through color coding based on the frequency of stem (region) occurrences in the analyzed set of structural conformations generated by the RNA folding programs. 2D stem histograms can be explored in conjunction with previously drawn structures and stem traces (discussed below). The connection between a particular region in a histogram and the same region in the current structure drawing is labeled on portions of the structure corresponding to the symbolic stem representation. The cumulative nature of the data stored in the stem histogram is ideally suited to drawing composite or *consensus structures*. These are built out of regions (stems) with frequencies between 50% and 100%. The user can select specific thresholds.

The Stem Trace provides a view orthogonal to that of the 2D Stem Histogram. Every unique triplet of 5' start position, 3' stop position, and stem length is plotted on the vertical axis. The horizontal axis denotes the generation number, for the GA-generated files, or suboptimal solution space ranking, for DPA results. In this way the most persistent stems and structural motifs in a given solution space form the most continuous lines. The stem traces are also color coded, based on frequency of occurrence or stem energy (on a normalized scale) as the user's coding criterion.

Another way to view a stem trace is based on the fact that a vertical "slice" cutting 'through a trace at any given point contains information equivalent to the region table for an entire structure. From this information, extracted from a trace with a click of a mouse button, a structure can be drawn without any need for file I/O. As a result of this data representation and ease of extraction, a series of structural drawings based on the data extracted from a selected stem trace range can be displayed in a slow-motion movielike sequence.

This function is particularly useful in the analysis of the GA folding pathways, based on the data written out by the GA at every generation, without the need to run the program itself. Structurelab provides a set of functions related to the stem trace plots that perform stem information display, stem and structure energy calculations, structure drawing, and the previously mentioned labeling of the related structure drawings.

11.4.4 WWW Interface

A WWW/http homepage for access to the system as well as an automatic demo run of the GA is available. The homepage (html code) gathers initial arguments (see figure 11.10) and invokes a WWW server (C code), which in turn starts the system. The user has to specify the name of the display, that is, the host name and the domain name of the user's machine (such as `fcsparc1.ncifcrf.gov` or `fcsparc1`, as shown in figure 11.10) and choose starting the entire system or the "canned" GA demonstration. An X-terminal window brought up at the indicated user's host display is utilized as the main Structurelab I/O window, with the regular menus and graphics windows providing the graphical interface.

The GA demo requires the absolute minimum of interaction from the user while it illustrates a live run of the GA on the MasPar coupled with Structurelab's drawing program running on the central controller machine. The Structurelab Workbench choice allows the user to explore all the tools of the system within limits of a predefined network of machines. At the moment access to these demos is limited to password holders.

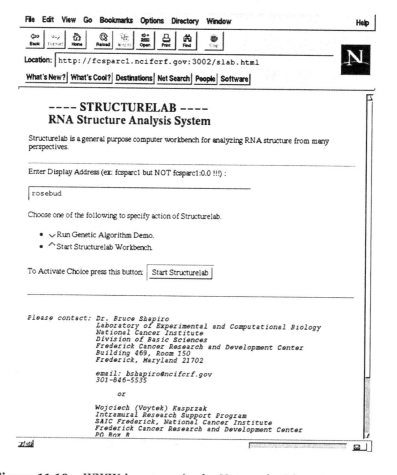

Figure 11.10 WWW homepage (under Netscape) with options to run a GA demo or the entire system.

11.5   Structurelab in Use

In our laboratory Structurelab has been used successfully to derive RNA secondary structures that are supported by biological data and agree with currently accepted models. For example, using the GA results with the system's analysis tools, we were able to arrive at a secondary structure model for the 5' NCR (noncoding region) of poliovirus 3 very similar to the one predicted by Skinner et al. (1989). All of the known structures shown to be biologically significant are present in the structure as well as some stems not present in the model but for which support can be found (Currey and Shapiro, 1997).

The structure for the 5' NCR of poliovirus 3 was derived by collecting the lists of stems produced by the genetic algorithm. The GA was run 100 times, collecting the most optimal solutions produced based on free energy. The lists of stems from these 100 solutions were plotted using the 2D Stem Histogram to create a stem histogram graph and to output a list of stems according to a specified threshold. For comparison purposes the structure for the 5' NCR of poliovirus 3 was also predicted using Mfold. Lists of stems were collected using the optimal plus 9,999 suboptimal solutions. The lists of stems from these 10,000 solutions were analyzed in the same way. Above a 55% frequency threshold, the GA found 18 of the 21 stems present in the model, including all 12 of the stems predicted by the DPA. Also at the 55% threshold, the GA predicted only 11 stems not found in the model while the DPA predicted 24 stems not found in the model. Threshold was set at 55% to allow as many stems as possible to be selected but to eliminate competing or overlapping structures. Thus, the 2D Stem Histogram allowed us to determine a structure by selecting frequently recurring stems and to successfully compare a new methodology to established methods of RNA structure analysis. Structurelab also incorporates the Significance/Stability Test, which, when applied to the 5' NCR of poliovirus 3, indicated regions of significant structure that correlated with known structural elements.

The list of stems produced by the 2D Stem Histogram determines the composite structure, but it is often necessary to visualize it, its substructures, or elements within it. We were then able to draw the overall structure and various structural elements of the 5' NCR of poliovirus 3 and compare it with the known structure by using Structurelab's visualization capabilities. The structure visualization functions were able to illustrate well those structural elements important in replication and in neurovirulence of the poliovirus 5' NCR. This approach is delineated in the flow chart under single sequence data (see figure 11.11).

Multiple sequence data can also be analyzed with Structurelab. Figure 11.11 shows some of the tools that are applicable to analyzing multiple sequences. It is particularly useful for finding common structural elements among related sequences. For example, poliovirus 1 Mahoney and poliovirus 2 Lansing were folded with the GA using the same parameters as for poliovirus 3 Leon above. The outputs were merged, an average threshold

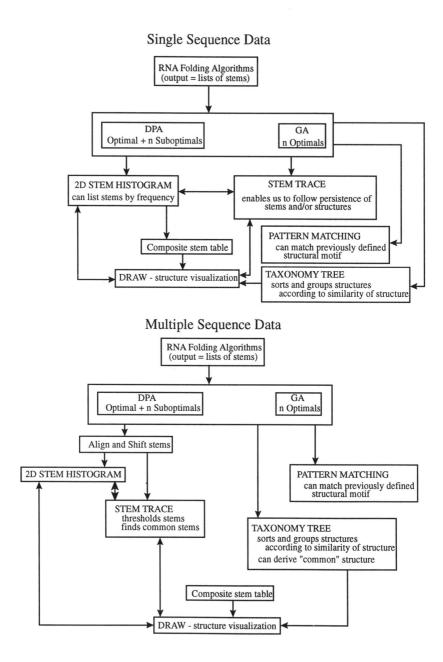

Figure 11.11 Schematic demonstrating some of the frequently used tools in Structurelab for both single sequence analysis and multiple sequence analysis. This diagram shows how these tools are interconnected and can be used.

per sequence of 50% was applied, and a "consensus" structure for poliovirus was derived (these three wild-type strains of poliovirus are closely related). This structure for the polioviruses compared more favorably to the accepted structure than the individual structures predicted for poliovirus 1 and 2. Additionally, all three serotypes of poliovirus were folded with the various parameters in the GA adjusted to allow a GU base pair to close a stem, but also to allow a one-base-pair peel-back if the base pair closing the stem was GU. Outputs from these runs were merged and analyzed in the same fashion by selecting stems occurring at an equivalent threshold of 50% per sequence. Both of these results were compared to the conserved stems found in the 10 polioviruses using Mfold (Le and Zuker, 1990). The "consensus" structure produced for poliovirus with the GA found 12 of the 21 stems in Skinner's structure and only seven stems not present in the model in the first case.

In the second case, 13 of the 21 stems present in Skinner's structure and only six stems not present in that structure were found. This is in contrast to the results obtained from Mfold when conserved stems derived from the optimal and numerous suboptimal structures were analyzed (Le and Zuker, 1990): 12 of the 21 stems reported by Skinner et al. (1989) were found, which is comparable, but 17 stems not in the accepted model as well as seven additional conflicting conserved stems were also predicted, making it difficult to extract the correct structure from the large group of stems. Both the 2D Stem Histogram and the Stem Trace were used to analyze the output of the folding programs. Because base positions of similar structural elements often differed in the poliovirus sequences, the 2D Stem Histogram was very helpful when we used the histogram graph portion to demonstrate structures by combining the three graphs into one display. In addition, Stem Trace can be used for a similar type of analysis whereby one can sort the stems both on a 5' basis and a 3' basis, and thus similar but shifted stems may be more easily discerned since they will be displayed physically close to one another. The interactive aspects of Stem Trace are particularly useful, since you can use the mouse to click on any stem or structure to display its boundaries and frequency. The frequency color mapping in both of these tools enabled us to determine at a glance the high-frequency stems and to determine visually the frequency of stems in which we were particularly interested. Stem Trace was useful in following given stems produced by multiple GA runs on the three strains of the poliovirus and was used to determine stems comprising the consensus structure. It was also helpful to see why some stems that we would have liked to have seen did not appear, or appeared at low frequencies. This occurred either because of other conflicting stems or because alternate structures outcompeted them. Using the energy calculation tools, we determined the energies of the alternate substructures and overall structures to better understand the dynamics of the folding pathways. The structure visualization functions can be activated using Stem Trace, as described above, and it was useful to select substructures to draw in the context of the entire structure to again better understand the dynamics of alternate structures.

As shown in the flow chart in figure 11.11, the use of these tools is fluid and at any point one may choose another tool to enhance the data. Additionally, the ease of use of the structure analysis tools is facilitated by the networking capabilities of the system since this aspect is transparent to the user. The outputs of the various analysis tools eventually converge and contribute to a composite stem table that can be used in the structural visualization package to draw a representative structure.

The success of Structurelab and the GA at determining the RNA secondary structure for poliovirus led us to apply these tools to the 5' NCR of coxsackievirus. Tu et al. (1995) reported that the cardiovirulent phenotype in CVB3 sequence is related to a single base change in the 5' NCR. In this work we showed that the single nucleotide change causes a potential alteration in the secondary structure involving this nucleotide. This was not apparent from the previously published consensus RNA secondary structure model for enteroviruses (Le and Zuker, 1990); rather, it was suggested by the GA and Structurelab methodology described here.

Although coxsackievirus and poliovirus are both enteroviruses and share many viral mechanisms, the virulence "switch" in poliovirus is primarily related to *translational* efficiency whereas the mechanism in coxsackievirus appears to be related to *transcriptional* efficiency. If nucleotide 234 is U, the virus is cardiovirulent; if nucleotide 234 is C, the virus is attenuated and transcription is markedly decreased. The studies demonstrated no role for the capsid or other regions to determine the virulence phenotype. Accepted models of the RNA secondary structure for the 5' NCR in enteroviruses have assumed the same structure for all enteroviruses including coxsackievirus and poliovirus. In the accepted models, the region containing the nucleotide responsible for attenuation of virulence in CVB3 is devoid of structure. However, recent studies utilizing the GA and Structurelab methodology place secondary structure elements in this region and demonstrate alteration of these structures with mutation of this nucleotide (Tu et al., 1995). The visual structure analysis tools were employed to select frequently recurring stems and arrive at this structure. The features unique to CVB3 occur in the region identified as determining cardiovirulence and affecting RNA transcriptional efficiency. Thus, a probable secondary structure map for CVB3 contains elements of the currently accepted structure and yet has unique features.

Since mutations at nucleotide position 234 with either A or G have not been found to date in wild-type CVB3s, we created two "mutant" sequences using the virulent CVB3/20 as the template. In one sequence A replaced U at nucleotide 234 (CVB3/20sa), and in the other sequence G replaced U at nucleotide 234 (CVB3/20sg). When these sequences were folded with the genetic algorithm, nucleotide 234 was involved in pairing much more frequently (approximately 80% of the time for each sequence). Furthermore, both sequences had nucleotide 234 involved in longer and thus more stable stems. Comparing the total pairing frequency, each had an individual stem above the 50% threshold, thus contributing to the overall secondary structure map. CVB3/20sa had nucleotide 234 involved in a stem five base pairs in length occurring at a threshold of 54%, and CVB3/20sg had nucleotide 234 involved in a stem eight base pairs in length at a threshold of 52%. The features of the structural analysis tools were used to determine the characteristics of the stems and their frequency, thus enabling this analysis. The GA predicted that G or A base substitution at position 234 contributed to enhanced stability of the secondary structure(s) in that region even more than C base substitution, which prompted the question of whether enhanced stability of these two variants correspond to decreased virulence and replication. Preliminary data from Dr. Steven Tracy's laboratory indicate that in a construct where G replaces U at nucleotide 234, replication is severely diminished, even more so than when C replaces U. The analysis tools of Structurelab were thus able to provide useful information to corroborate and guide biological experiments.

Using the multiple sequence data approach of Structurelab as described above, the structure analysis of the CVB3s was extended by applying the genetic algorithm to fold the entire 5' NCR of the CVB3 sequences. A proposed secondary structure for this entire region was generated by applying a 50% threshold to the predicted stems, which has multiple stems in common with both the model proposed by Skinner et al. (1989) for poliovirus and the enteroviruses and the global enterovirus model proposed by Le and Zuker (1990). However, there are unique features that occur primarily within the domain described by stem 195–434, and specifically the stems unique to the genetic algorithm model that includes stems involving nucleotide 234.

Structurelab has also proven to be a useful tool in several other collaborations. Some examples include the study of structural influence on translation efficiency of the LamB gene and the L11 operon of *E. coli* (Margalit et al., 1989). Work has also been done on refining the secondary structure of the Rev Responsive element sequence of HIV-1 (Dayton et al., 1992). Structurelab has been used to explore RNA termination sites (Cheng et al., 1991) and is currently being used in exploration of the structural aspects of the HIV dimerization site (Sakaguchi et al., 1993).

Structurelab has demonstrated that it is a powerful and flexible tool for analyzing RNA structure and for gaining insight into RNA structure-function relationships within the enteroviruses. It continues to be applied to other viral systems and has been productive in analyzing other sequence data, supporting and guiding other biological experiments. Use of the system reveals needs for new capabilities and refinement of the existing tools, thus guiding its further development.

11.6 Future Directions and Challenges

Although methodology for RNA structure prediction has continued to improve, supporting biological or biochemical data remains useful. Results are better supported when biological data confirm them or when phylogenetic comparison of related sequences lends credence. The biological data can confirm predicted structural elements and guide some structural analyses by their use within the programs. Conversely, predicted structures can suggest and guide further biological experiments. Used together, they become very powerful tools for understanding RNA structure-function relationships.

RNA structure determination, however, still is a very challenging problem. Difficulties arise from various perspectives all of which represent future challenges. A few of these are outlined below.

1. RNA secondary structure prediction requires more refined rules that take into account more of the contextual characteristics of the RNA environment. These include not only various morphologic structures described previously but also more complicated tertiary interactions. Some of these rules may be determinable with the discovery of some secondary and tertiary motifs.

2. Environmental factors such as solvent, temperature, ionic concentrations, and protein interactions contribute to the structure and function of RNA molecules. Any defined rules should take these environmental factors into account.

3. More experimental information would be very helpful. X-ray and nuclear magnetic resonance structures are starting to become available. This will add greatly to our knowledge base for the folding of the molecules. Subtleties that are not so readily discernible by standard biochemical techniques will become more apparent, thus improving the structure prediction capabilities. In addition, new techniques such as atomic force microscopy will help us to visualize the structures. The combination of visual feedback and computer prediction methods may be very helpful in elucidating structural information.

4. More computing power and better algorithms in conjunction with the above-mentioned items will also help in the structure determination problem. As additional information is added, especially at the atomic level, procedures such as molecular dynamics and mechanics become more feasible. These, however, are computationally very expensive to apply to a molecule of any size. Today's computers, for example, cannot do atomic-level computations on RNAs anywhere near the size of HIV ($\sim$9,500 nucleotides in length). An interesting challenge that, again, is nowhere near feasible today is the de novo folding of a molecule of this size. This means simulating a fold from a single strand to secondary and tertiary interactions taking into account the environmental factors.

5. Adding information from families of sequences will help ultimately determine structure. Again, this assumes that families have similar folding patterns even though they may differ somewhat in their primary sequence. Phylogenetic comparisons on the sequences and structural levels can provide important clues to the folding characteristics of the sequences. Known

structures of one sequence or even determining known motifs would help determine families of structures.

6. An added complication is the interaction of RNAs with other RNAs and/or proteins. Knowing the structure of a noncomplexed RNA may not be that helpful in determining the structure of the complexed RNA. Structural changes can take place in an RNA, as well as a protein, as a function of complex formation. De novo determination of such complexed structures again would require either significant computing power and/or experimental data.

Systems like Structurelab described here as well as other algorithmic techniques help in the ultimate goal of RNA structure-function determination. However, we are still a long way from treating problems such as this in a black box manner. More experimentation, more interdisciplinary collaborations, and more understanding of the chemistry and physics of molecular interactions as well as breakthroughs in combinatoric algorithms and computer power will ultimately help to unravel this challenging problem.

Acknowledgments

We thank Dr. Hugo Martinez for providing his rna_2d3d package, and John Owens for creating some of the figures in this chapter. We extend our gratitude to people of the Franz Inc. for their help and support. In addition, we thank Dr. Jacob Maizel, Jr., for his support of this project.

Glossary

active site. The part of a protein (or ribozyme) involved in catalyzing a reaction, usually a binding site for a reaction intermediate, which provides a low-energy path between two stable states (chapter 10).

agglomerative. A clustering algorithm that successively refines the solution, starting with one in which each instance in the data set is a cluster (chapter 5).

Alu sequences. The most abundant class of interspersed repetitive sequences in human genomes. A human genome is believed to contain on the order of one million of copies of Alu sequence variants. Recognizing these Alu sequences can help in gene discovery in computational biology (chapters 1 and 4).

amino acid mutation (substitution) matrix. A scoring matrix used for sequence comparison. For standard sequence similarity searches, PAM and BLOSUM series of matrices have been constructed by counting the frequency of amino acid mutations in known proteins. A mutation occurring frequently is given a higher similarity (lower distance) score than an infrequently occurring mutation. When searching for structural similarities, matrices need to be derived from a set of proteins with known 3D structures (chapters 2, 6, and 10).

amino acid residues. The monomeric units of a protein. Proteins are composed of 20 amino acids that are linked together by peptide bonds. The amino acids and their abbreviations are alanine (A), cysteine (C), aspartic acid (D), glutamic acid (E), phenylalanine (F), glycine (G), histidine (H), isoleucine (I), lysine (K), leucine (L), methionine (M), asparagine (N), proline (P), glutamine (Q), arginine (R), serine (S), threonine (T), valine (V), tryptophan (W), and tyrosine (Y). The residues are usually classified by their properties, for example, charged, polar, or hydrophobic. The primary structure of a protein consists of a sequence of amino acids (chapters 2, 3, 5, 6, and 10).

annealing mutation. A process used in genetic algorithm for RNA folding whereby the probability of a mutation in an RNA secondary structure decreases with each generation by a defined schedule. Therefore, more mutations occur in the beginning of a run than at the end (chapter 11).

block. Region of local similarity in a multiple alignment of a group of protein sequences. Blocks are generally ungapped; no insertions or deletions are allowed in the local multiple alignment (chapter 2).

capsid. A protein coat of a viral particle. In the context of this book, it represents the portion of the viral RNA that encodes these proteins (chapter 11).

cDNA. Complementary DNA, the sequence of an RNA molecule, usually mRNA, copied into DNA (which is more stable than RNA) by reverse transcriptase. cDNA is used to study the mRNA expressed in a cell, since mRNA itself is not suitable for sequencing (chapter 10).

chromosome. A structure composed of a very long DNA molecule and associated proteins that carries part (or all) of the hereditary information of an organism. It is especially evident in plant and animal cells undergoing mitosis or meiosis, where each chromosome becomes condensed into a compact, readily visible thread (chapter 10).

classification. The activity of determining whether or not an unlabeled object belongs to an existing class (or family) (chapters 2, 4, 5, 6, and 10).

clone. A DNA molecule that has been replicated by means other than those related to its native cell cycle. To "clone a gene" means to produce many copies of a gene by repeated cycles of replication (chapter 1).

clustering. The organization of data so that related information is grouped together. There are several approaches to clustering, including numerical, statistical, and conceptual (chapters 5, 6, 7, 10, and 11).

codon. Triplet of contiguous bases in mRNA that code for specific amino acids, which in turn are used for building proteins; see also *translation* (chapter 11).

consensus sequence. Given a set S of k sequences of length N, a consensus sequence c is a possibly different sequence of length N where the jth entry in c is the one that appears most often in the jth position of all sequences in S. Consensus sequences are often used in protein and DNA analysis (chapter 3).

contig. An assembly of overlapping sequences into a single long sequence (chapter 10).

database. One or more large structured sets of persistent data, usually associated with software to update and query the data (chapter 10).

data mining. Analysis of data in a database using tools that look for trends or anomalies without knowledge of the meaning of the data except for some measure of similarity; see also *knowledge discovery* (chapters 5, 9, and 10).

data model. A formal language for describing data (e.g., as records consisting of some set of fields or as a graph with some description of edges) and for performing operations on the data (e.g., match, insert, delete, update, and count) (chapter 10).

decoding algorithm. An algorithm that takes a sequence of symbols in a fixed alphabet and converts them into a representation of an object. The representation of the object can be a sequence of symbols itself but typically is more verbose and, for humans, easier to read. For example, consider a DNA sequence AGTCAGTTTT. One could encode this as AGTC(1,3)(7,3). The pair (1,3) indicates that AGT should follow the initial AGTC. The pair (7,3) indicates that TTT should follow AGTCAGT.

The decoding algorithm consists of the following two steps: (1) replace each pointer by a sequence of pointers to individual letters, and (2) replace the new pointers by their targets in the left-to-right order. Thus, in our example, the first step would yield AGTC(1,1)(2,1)(3,1)(7,1)(8,1)(9,1) and the second step would yield the original sequence (chapter 1).

distributed system. A collection of processing devices, which may or may not be identical, used to solve some sets of problems (chapter 10).

divisive. A clustering algorithm that successively refines the solution, starting with one in which all instances in a data set belong to a single cluster (chapter 5).

DNA. Deoxyribonucleic acid, a long molecule shaped like a twisted rope consisting usually of two complementary strands forming a double helix. The strands are held together by hydrogen-bonded nucleotides containing the bases adenine (A), cytosine (C), guanine (G), and thymine (T). In its double helix form, A pairs with T and G with C. Genetic information is stored as the sequence of the bases. Information is transmitted by the pairing of bases on one strand of the helix with complementary bases on the other strand; see also *nucleotide* (chapters 1, 4, 9, and 10).

dynamic programming algorithm (DPA). A programming methodology commonly used for optimization whereby current optimal solutions in the algorithm are constructed from previous optimal subproblem solutions. One of the major algorithms used for RNA folding is based on this paradigm (chapter 11).

encoding algorithm. An algorithm that takes an object and converts it into a sequence of symbols according to an encoding scheme. The original object itself may be represented by a sequence of symbols (chapter 1).

encoding length. The number of symbols needed to encode an object by a particular encoding algorithm. Typically, binary symbols "0" and "1" are assumed (chapter 1).

encoding scheme. A precisely defined method for representing objects as sequences of symbols in a fixed alphabet. Part of the definition includes the specification of a decoding algorithm (chapter 1).

enteroviruses. RNA viruses found in the human intestine (chapter 11).

evolutionary tree. A tree structure containing observed organisms or DNA sequences in the leaf nodes, and putative ancestors in the internal nodes (chapter 1).

exon. Segment of a eukaryotic gene that consists of DNA coding for a sequence of nucleotides in mRNA. An exon can encode amino acids in a protein; it is usually adjacent to a noncoding DNA segment called an intron (chapter 10).

exploration dag (E-dag). A directed acyclic graph (dag) structure for expressing computation models that follow a generate-and-test paradigm—generate a pattern and then test whether it is "good." Each node of an E-dag represents a candidate pattern, and directed edges between nodes represent interdependencies among patterns. An E-dag traversal visits a node if and only if all nodes that point to that node have been tested and are "good." Therefore, every exploration order generated by an E-dag traversal defines an optimal execution (chapter 8).

expressed sequence tag (EST). Partial sequence of a cDNA (which in turn represents an mRNA). ESTs can be used to "tag" genes expressed (i.e., used to make proteins) in specific cell types (chapter 10).

gapped fingerprint. A possibly noncontiguous subsequence of a segment *Seg* that begins with the segment's first letter. A gap at position p means that when forming the fingerprint, one does not include the letter at position p in *Seg*. For example, "ATT" is a fingerprint of a segment "ACGTTG" with two gaps (one gap at position 2 and one gap at position 3) (chapter 4).

GenBank. A repository for DNA, RNA, and protein sequences with annotated information, available for computer searches and sequence extraction. Users can access it on the web at http://www.ncbi.nlm.nih.gov/ (chapters 3, 10, and 11).

gene. A hereditary unit within a genome corresponding to a contiguous stretch of DNA. Originally, it referred to the region of DNA controlling inheritance of a trait, usually containing a sequence coding for a single protein. Now we know that some "genes" code for nested or partially overlapping sets of proteins with related functions, and some "genes" only partly control the inheritance of a trait, so the precise meaning depends on context (chapters 1, 10, and 11).

gene expression. Phenomena related to the ultimate biological function of a gene. The most prominent step in gene expression is the transcription of DNA into RNA. The abundance of RNA transcripts for a particular gene is usually referred to as the transcription level or expression level (chapter 1).

genetic algorithm (GA). A nondeterministic programming methodology commonly used for optimization that borrows from concepts in biological evolution. It uses concepts such as genetic mutation and crossover coupled with survival of the fittest to determine surviving members of a population of elements to be optimized. Several generations of this process are required to determine a potential optimum (chapter 11).

genetics. The science of genes and heredity, through the study of family trees and experiments with model organisms (chapter 10).

genome. The complete set of DNA (or RNA, for retroviruses) for an organism. The genome is partitioned into one or more chromosomes (chapters 1, 3, and 10).

genomic DNA. The cell's hereditary DNA, for example, chromosomal DNA, including regulatory regions and introns normally spliced out in mRNA, as opposed to cDNA made from mRNA (chapter 10).

graph isomorphism. A graph G_1, defined by vertices V_1 and edges E_1, is isomorphic to graph G_2, defined by vertices V_2 and edges E_2, if there is a one-to-one function mapping vertices of G_1 onto vertices of G_2, $f : V_1 \rightarrow V_2$ satisfying the property that $\{u, v\} \in E_1$ if and only if $\{f(u), f(v)\} \in E_2$ (chapter 9).

graph substructure. A connected graph embedded in another graph (chapter 9).

hidden Markov model (HMM). A probabilistic generative model for sequences, for example, for the members of a protein family. An HMM is a graph where nodes are called states and edges are called transitions. Each state is labeled with a probability distribution over the alphabet of the sequences to be modeled, and there is a probability distribution over the transitions out of each state. The probability of a particular sequence, given a fixed path through an HMM, is the product of the transition probabilities along the path and the probability in each state of the corresponding letter in the sequence. The total probability of a sequence given an HMM is the sum over all paths through the HMM of the probability of the sequence given the paths (chapters 1, 2, 3, 5, and 10).

high-complexity DNA sequences. DNA sequence data that are not of low complexity. Normally, such sequences contain no repetitions; compare *low-complexity DNA sequences* (chapter 1).

homologs. Proteins derived through evolution from a common ancestor protein. Homologs usually have similar general functions, such as metabolizing proteins or RNA, oxidizing alcohol groups, and so forth. Homologous proteins typically have similar 3D structures, a property that is conserved better than amino acid sequence during evolution. Homologs typically have one or more shared motifs because motifs are usually in structurally important segments and thus well conserved. Similarly, motif order and spacing, properties of their 3D structure, are conserved in evolutionarily distant proteins. Types of homologs include orthologs and paralogs. The term "homologous" is often misused to mean "similar," though two sequences can be similar to some degree without being homologous (e.g., convergent evolution) (chapters 3 and 10).

HTML. Hypertext Markup Language, used to create World Wide Web pages and links among them (chapter 11).

hybridization experiment. A biochemical experiment where DNA or RNA probes are mixed with DNA or RNA samples under conditions that permit the molecules to anneal if there is sufficient complementarity between the probes and the samples. For example, a short oligomer may hybridize to a longer clone if the longer clone contains a sequence complementary to that of the oligomer (chapter 1).

incremental clustering. A clustering technique for which instances are considered sequentially (chapter 5).

interoperability. The ability of software and hardware on multiple machines from multiple vendors to communicate (chapter 10).

intron. Noncoding region of a eukaryotic gene that is transcribed into an RNA molecule but is then excised by RNA splicing when mRNA is produced; compare *exon* (chapter 10).

inverse protein folding. The process of fitting a known structure to a given sequence (rather than the more standard practice of trying to predict the structure of a protein from the sequence) (chapter 5).

knowledge discovery. A field of artificial intelligence that provides computational tools for extracting useful information from datasets; see also *data mining* (chapters 5, 9, and 10).

linear time algorithm. An algorithm whose running time is linearly proportional to the size of its input (chapter 1).

low-complexity DNA sequences. DNA sequences that contain patterns, typically in the form of biased sequence composition (preponderance of particular bases) or simple tandem repetitions; compare *high-complexity DNA sequences* (chapter 1).

minimum description length (MDL) principle. A criterion for identifying the best theory to describe a set of data. Let $L(T)$ denote the number of bits needed to encode the theory T, and let $L(D|T)$ denote the number of bits needed to encode the data D with respect to T. $L(T)$ measures the complexity of the theory, and $L(D|T)$ measures how well the theory matches the data—the fewer bits the better. The MDL principle states that the best theory is the one with the least number of bits required to encode the theory and the data. That is, the best theory T is the one that minimizes $L(T) + L(D|T)$ (chapters 1 and 9).

motif. A pattern that occurs often (at least approximately) and may have some bearing on functionality. For example, in a protein, a motif consisting of a subsequence of consecutive amino acids may correspond to sites of structural or functional importance that tend to be highly conserved by evolution. In a DNA sequence, a motif may be a regulatory signal whose appearances are recognized by a DNA-binding enzyme. An open question is whether or not motifs should be consecutive subsequences and whether a probabilistic description might be better (chapters 3, 4, 5, 6, 8, 10, and 11).

motif as probability model. Assigns a certain probability to each subsequence of length N. This probability is the degree to which the subsequence is an appearance (also called an instance) of the motif model, which concretely is a matrix of probabilities. The number of columns of the matrix is the length N of the motif. The number of rows is the size of the alphabet: 4 for DNA motifs and 20 for protein motifs. The entry of the matrix at row number i and column number j is the probability of finding the letter numbered i at the jth position in a subsequence that is an instance of the motif. The motif matrix can be thought of as the frequencies that would be observed in each column of a multiple alignment of all possible examples of the pattern (chapter 3).

MP-score. Mean sum of pairs score for a column in a multiple alignment. It is the SP-score divided by the total number of possible pairs of amino acids; compare *SP-score* (chapter 2).

multiple instruction multiple data (MIMD). A parallel computer paradigm in which a collection of processors run mostly independently of one another in parallel. The processors are usually connected by a fast communications network to pass data back and forth, when communication is necessary (chapter 11).

network file system. A file system that may be accessible from several different computer architectures but appears as if it is associated directly with the given workstation. The files may actually reside on totally separate disks (chapter 11).

noncoding region (NCR). The portion of the genome sequence that does not code for protein(s). In contrast to introns, which occur in DNA or mRNA precursors, NCRs occur in mature mRNA (chapter 11).

nucleotide. One of the building blocks of DNA and RNA; one of four heterocyclic bases attached to a sugar, (deoxy)ribose. The sugar is linked to phosphates to form the helical backbone of DNA. The four nucleotides in DNA are adenine (A), cytosine (C), guanine (G), and thymine (T). In RNA, thymine is replaced by uracil (U). Each nucleotide complements one other nucleotide by hydrogen bonding: A pairs with T (or with U in RNA) and C pairs with G (chapters 10 and 11).

object-oriented design. A design method in which a system is modeled as a collection of cooperating objects and individual objects are treated as instances of a class within a class hierarchy. Design usually consists of at least four stages: identify the classes and objects, identify their semantics, identify their relationships, and specify class and object interfaces and implementation (chapter 10).

oligomer. A molecule consisting of a sequence of structural units. In the context of DNA, an oligomer (oligonucleotide) is a sequence of nucleotides (chapter 1).

orthologs. Homologous genes in different species; divergence was due to speciation, and the function of the orthologs is generally conserved (chapter 10).

paralogs. Genes that arose by duplication and subsequent divergence of a gene within a genome. Such a redundant gene may evolve a new (often related) function (chapter 10).

Persistent Linda (PLinda). A software system for running robust distributed parallel programs on networks of workstations that are occasionally idle. It can be used for any scientific data-mining application without disturbing workstation owners. An application written in PLinda programs "retreats" when a workstation is being used by any other application, so users do not notice its presence, but the PLinda environment ensures that the computation completes correctly and effectively.

The parallel execution model of PLinda is based on Linda, developed by Dave Gelernter and Nick Carriero at Yale University, which provides a set of language constructs that can be added to any programming language to facilitate writing parallel programs in that language. PLinda adds a set of extensions such as "lightweight transactions" and "continuation committing" to Linda to support fault-tolerant parallel computations on networks of intermittently idle, heterogeneous workstations. PLinda models nonidleness (e.g., user returns to workstation) as failure. In this way, PLinda programs can run in a manner that is fault tolerant and does not disturb the owners of the workstations at all (chapter 8).

phosphodiester linkages. Part of the DNA and RNA consisting of phosphate groups that link sugars together to form the backbone of the molecule (chapter 11).

position-specific scoring matrix (PSSM). Given a block described in chapter 2, a PSSM is a matrix of numbers representing the occurrence of each amino acid in each position of the alignment. The PSSM is as wide as the block from which it is derived and has one row for each amino acid. It is used to scan protein sequences for matches to the block (chapter 2).

primary DNA sequence. The sequence of ordered bases adenine (A), cytosine (C), guanine (G), and thymine (T) in DNA. None of the 2D or 3D characteristics of DNA are considered; only the sequential ordering is considered (chapters 1, 4, and 10).

primary protein sequence. The sequence of amino acids in protein. None of the 2D or 3D characteristics of protein are considered; only the sequential ordering is considered (chapters 2, 3, 4, and 10).

primary RNA sequence. The sequence of ordered bases adenine (A), cytosine (C), guanine (G), and uracil (U) in RNA. None of the 2D or 3D characteristics of RNA are considered; only the sequential ordering is considered (chapter 11).

profile. A sequence pattern specified by the probability of observing each type of residue (nucleotide or amino acid) at each position in all instances of a motif (chapters 3 and 10).

protein. A polymer of amino acids, usually linear except for disulfide crosslinking. A protein's sequence, or primary structure, generally determines its local and global 3D shapes (secondary and tertiary structure), which determine its chemical properties. Proteins with similar sequences are likely to have similar structure and function. Classes of protein include enzymes, which catalyze most of the chemical reactions in the cell; structural proteins; receptors; and proteins that transport and store small molecules (e.g., hemoglobin) (chapters 2, 3, 4, 5, 6, and 10).

protein family. A set of homologous proteins from multiple species. Typically they have similar structures and functions. Protein families can be classified into subfamilies and superfamilies (chapters 2, 3, 4, 5, 6, and 10).

protein secondary structure. The local conformation of the protein backbone. The most common folding patterns are helices, sheets, and turns (chapters 5 and 6).

protein supersecondary structure. Recurring groupings of secondary structure units (e.g., $\alpha\alpha$-unit, $\beta\alpha\beta$-unit, β-barrel) (chapters 5 and 6).

protein tertiary structure. The global conformation of a protein folded in 3D space (chapters 5 and 6).

pseudoknot. A tertiary interaction of an RNA that involves base pairings between loop structures and/or loop structures and free base regions adhering to certain constraints (chapter 11).

replication. Duplication of RNA or DNA from its template (chapter 11).

retrovirus. A form of RNA virus that transforms its RNA into DNA, which is incorporated into the infected cell's nucleus. Viral products are then generated using the cell's genetic machinery (chapter 11).

reverse transcription. Making cDNA from mRNA, done by retroviruses that store their genetic material as RNA and by molecular biologists making expressed sequence tags (chapter 11).

ribose sugar. Five-carbon sugar forming part of the DNA (deoxyribose) or RNA (ribose) backbone that ultimately forms a long chain. Bases are linked to these sugars (chapter 11).

ribosome. A spherical particle consisting of ribosomal RNA and proteins on which protein synthesis takes place (chapter 11).

RNA. Ribonucleic acid, a long molecule similar to DNA but usually single stranded, except when it folds back on itself. It differs chemically from DNA by containing ribose sugar instead of deoxyribose and containing the base uracil instead of thymine. Thus, the four bases in RNA are adenine (A), cytosine (C), guanine (G), and uracil (U). mRNA (messenger RNA) may have parts spliced out (usually in eukaryotes); the remaining sequence contains instructions for protein synthesis. Other forms of RNA include tRNA (transfer RNA) and rRNA (ribosomal RNA), involved in protein synthesis; and snRNP (small nuclear ribonucleo proteins), which are complexes of RNA and protein involved in RNA splicing and other nuclear functions; see also *nucleotide* (chapters 8 and 11).

RNA consensus structures. RNA structures or motifs that are in common among various sequences usually within the same family (chapter 11).

RNA region/stem. Helical base-paired contiguous segment of hydrogen-bonded nucleotides, represented in Structurelab (described in chapter 11) by a quadruplet consisting of the starting base position of the segment, the ending base position of the segment, the size of the helical segment, and its energy (chapter 11).

RNA secondary structures. Single- and double-stranded structures that form in RNA when a single strand folds back on itself (chapters 8 and 11).

RNA tertiary interactions. Interactions that occur between unpaired bases of secondary structures of RNA. Sometimes tertiary interactions may occur with triple base pairing (chapter 11).

sequence alignment. Juxtaposition of two sequences that indicates a set of edit operations that transforms one sequence into the other. Specifically, aligning two homologous protein sequences means identifying corresponding amino acids in each sequence. Corresponding amino acids either have been left unchanged by evolution or have been derived by mutation from the same predecessor amino acid. A gap in an aligned sequence corresponds to an amino acid in the other sequence that arose through an insertion or whose matching amino acid was deleted by evolution in the first sequence (chapters 1, 2, 3, 4, 5, 6, 10, and 11).

sequence segment. A subsequence made up of consecutive letters, also called a fragment. For example, "DPM" is a segment of "YDPMNW" (chapter 4).

serotype. Strain of a virus causing a specific, independent immune response (chapter 11).

single instruction multiple data (SIMD). A parallel computer paradigm whereby a controller issues one instruction at a time, which is executed in parallel by many processors. These processors usually have access to their own local memory but may communicate with other processors via tightly coupled mesh communications, routers, or both (chapter 11).

splicing. Taking out noncoding parts of an mRNA (i.e., introns) and joining the remaining pieces (i.e., exons) to make a shorter mRNA in which all the codons are adjacent (chapters 10 and 11).

SP-score. Sum of pairs score for a column in a multiple alignment. It is the sum of scores from an amino acid substitution matrix for all possible pairs of amino acids in the column; compare *MP-score* (chapter 2).

string edit distance. The minimal weighted number of edit operations used to transform one string (e.g., DNA sequence) to the other. The edit operations include insert a letter, delete a letter, and change one letter to another. The weighting may depend on the application, though often a weight of 1 is given for each edit operation (chapters 1 and 4).

structural motif. A characteristic local structural pattern in the 3D structure of a protein. In general, the structural motif represents a building block of the 3D structure, but it is often related to a functional site of a protein. Sometimes the term "structural motif" specifically refers to a recurring spatial pattern of secondary structure elements. Examples include the helix-turn-helix motif and the β-α-β motif (chapters 5 and 6).

suffix tree. A data structure (related a special kind of tree called a trie) that compactly represents a string by collapsing a series of nodes having one child to a single node whose parent edge is associated with a string (chapter 4).

sugar-phosphate backbone. The combination of covalently linked phosphates and sugars that form the backbone of DNA and RNA. Covalently linked to this backbone are the bases (chapter 11).

supervised learning. Discovering a pattern that is shared by all the positive examples but by none of the negative examples from a given set of training examples labeled "positive" and another set of training examples labeled "negative" (chapter 3).

transcription. Making an RNA copy of a DNA sequence, by base pairing with the complementary strand (chapter 11).

translation. Making protein from the coding sequence on mRNA (codon), by reading three bases in the mRNA for each amino acid (chapter 11).

tree edit distance. The minimal weighted number of edit operations used to transform one tree to the other. The edit operations include insert a node, delete a node, and change the label of a node. The weighting can be related to the position in the tree or the kind of editing operation (chapter 8).

uniquely decodable scheme. An encoding scheme whose encoding function E and decoding function D are inverses, such that $o = D[E(o)]$. Such a scheme is therefore information preserving (chapter 1).

unsupervised learning. Discovering one or more statistically surprising patterns shared by one or more subsets of a given set of training examples. Unsupervised learning is different from supervised learning in that no "teacher" has classified training examples as "positive" or "negative." For each pattern that an unsupervised learner discovers, the learner also discovers which are the positive examples of this pattern (chapter 3).

References

Abiteboul, S., D. Quass, J. McHugh, J. Widom, and J. L. Wiener. 1997. "The lorel query language for semistructured data." *International Journal on Digital Libraries* 1(1):68–88.

Abola, E. E., F. C. Bernstein, S. H. Bryant, T. F. Koetzle, and J. Weng. 1987. "Protein Data Bank." In *Crystallographic Databases—Information Content, Software Systems, and Scientific Applications*, pp. 107–132 (Bonn: Data Commission of the International Union of Crystallography).

Abrahams, J. P., M. van den Berg, E. van Batenburg, and C. Pleij. 1990. "Prediction of RNA secondary structure, including pseudoknotting, by computer simulation." *Nucleic Acids Research* 18:3035–3044.

Agrawal, R., H. Mannila, R. Srikant, H. Toivonen, and A. I. Verkamo. 1996. "Fast discovery of association rules." In *Advances in Knowledge Discovery and Data Mining*, ed. U. M. Fayyad, G. Piatetsky-Shapiro, P. Smyth, and R. Uthurusamy, pp. 307–328 (Cambridge, Mass.: MIT Press).

Altschul, S. F., W. Gish, W. Miller, E. W. Myers, and D. J. Lipman. 1990. "Basic local alignment search tool." *Journal of Molecular Biology* 215:403–410.

Altschul, S. F., M. Boguski, W. Gish, and J. C. Wootton. 1994. "Issues in searching molecular sequence databases." *Nature Genetics* 6:119–129.

Anderson, B., and D. Shasha. 1992. "Persistent Linda: Linda + transactions + query processing." In *Research Directions in High-Level Parallel Programming Languages*, ed. J. P. Banatre and D. Le Metayer, pp. 93–109 (Heidelberg: Springer-Verlag Lecture Notes in Computer Science).

Apple Computer, Inc. 1997. "Interapplication communication." Available at http://devworld.apple.com/dev/techsupport/insidemac/IAC/IAC-2.html.

Atkinson, M., F. Bancihon, D. J. DeWitt, K. Dittrich, D. Maier, and S. Zdonik. 1990. "The object-oriented database manifesto." In *Proceedings of the Conference on Deductive and Object-Oriented Databases*, pp. 40–57 (New York: Elsevier Science).

Attwood, T. K., and M. E. Beck. 1994. "PRINTS—A protein motif fingerprint data base." *Protein Engineering* 7:841–848.

Attwood, T. K., and J. B. Findlay. 1993. "Design of a discriminating fingerprint for G-protein-coupled receptors." *Protein Engineering* 6:167–176.

Bacon, D. J., and W. J. Anderson. 1986. "Multiple sequence alignment." *Journal of Molecular Biology* 191:153–161.

Bailey, T. L., and C. P. Elkan. 1994. "Fitting a mixture model by expectation maximization to discover motifs in biopolymers." In *Proceedings of the Second International Conference on Intelligent Systems for Molecular Biology*, pp. 28–36 (Menlo Park, Calif.: AAAI Press).

Bailey, T. L., and C. P. Elkan. 1995. "The value of prior knowledge in discovering motifs with MEME." In *Proceedings of the Third International Conference on Intelligent Systems for Molecular Biology*, pp. 21–29 (Menlo Park, Calif.: AAAI Press).

Bailey, T. L., M. Baker, and C. Elkan. 1997. "An artificial intelligence approach to motif discovery in protein sequences: Application to steroid dehydrogenases." *Journal of Steroids, Biochemistry, and Molecular Biology* 62:29–44.

Bains, W. 1986. "The multiple origins of human Alu sequences." *Journal of Molecular Evolution* 23:189–199.

Bairoch, A. 1992. "PROSITE: a dictionary of sites and patterns in proteins." *Nucleic Acids Research* 20:2013–2018.

Bairoch, A. 1993. "The PROSITE dictionary of sites and patterns in proteins, its current status." *Nucleic Acids Research* 21:3097–3103.

Bairoch, A. 1994. "The SWISS-PROT protein sequence data bank: Current status." *Nucleic Acids Research* 22:3578–3580.

Bairoch, A., and B. Boeckmann. 1992. "The SWISS-PROT protein sequence data bank." *Nucleic Acids Research* 20:2019–2022.

Bairoch, A., and P. Bucher. 1994. "PROSITE: Recent developments." *Nucleic Acids Research* 22:3583–3589.

Baker, M. E. 1991. "Genealogy of regulation of human sex and adrenal function, prostaglandin action, snapdragon and petunia flower colors, antibiotics, and nitrogen fixation: Functional diversity from two ancestral dehydrogenases." *Steroids* 56:354–360.

Baker, M. E. 1994a. "*Myxococcus xanthus* C-factor, a morphogenetic paracrine signal, is similar to *Escherichia coli* 3-oxoacyl-[acyl-carrier-protein] reductase and human 17β-hydroxysteroid dehydrogenase." *Biochemical Journal* 301:311–312.

Baker, M. E. 1994b. "Protochlorophyllide reductase is homologous to human carbonyl reductase and pig 20β-hydroxysteroid dehydrogenase." *Biochemical Journal* 300:605–607.

Baker, M. E. 1994c. "Sequence analysis of steroid- and prostaglandin-metabolizing enzymes: Application to understanding catalysis." *Steroids* 59:248–258.

Baker, M. E. 1995. "Enoyl-acyl-carrier-protein reductase and *Mycobacterium tuberculosis* InhA do not conserve the Tyr-Xaa-Xaa-Xaa-Lys motif in mammalian 11β- and 17β-hydroxysteroid dehydrogenases and *Drosophila* alcohol dehydrogenase." *Biochemical Journal* 309:1029–1030.

Barrett, C., R. Hughey, and K. Karplus. 1997. "Scoring hidden Markov models." *Computer Applications in the Biosciences* 13(2):191–199.

Batzer, M. A., P. L. Deininger, U. Hellmann-Blumberg, J. Jurka, D. Labuda, C. Rubin, C. W. Scmid, E. Zietkiewicz, and E. Zuckerkandl. 1996. "Standardized nomenclature for Alu repeats." *Journal of Molecular Evolution* 42:3–6.

Baxter, R. A., and J. J. Oliver. 1994. "MDL and MML: Similarities and differences (introduction of minimum encoding inference—part III)." Technical Report 207, Department of Computer Science, Monash University, Clayton, Victoria 3168, Australia.

Benson, D. A., M. Boguski, D. J. Lipman, and J. Ostell. 1996. "GenBank." *Nucleic Acids Research* 24:1–5.

Benson, G., and M. S. Waterman. 1994. "A method for fast database search for all *k*-nucleotide repeats." *Nucleic Acids Research* 22:4828–4836.

Berg, O. G., and P. H. von Hippel. 1987. "Selection of DNA binding sites by regulatory proteins: Statistical-mechanical theory and application to operators and promoters." *Journal of Molecular Biology* 193:723–750.

Bernstein, F. C., T. F. Koetzle, G. J. B. Williams, E. F. Meyer, Jr., M. D. Brice, J. R. Rodgers, O. Kennard, T. Shimanouchi, and M. Tasumi. 1977. "The Protein Data Bank: A computer-based archival file for macromolecular structures." *Journal of Molecular Biology* 112:535–542.

Bleasby, A. J., and J. C. Wootton. 1990. "Construction of validated, non-redundant composite protein sequence databases." *Protein Engineering* 3:153–159.

Blumer, A., A. Ehrenfeucht, D. Haussler, and M. K. Warmuth. 1987. "Occam's razor." *Information Processing Letters* 24:377–380.

Borgida, A., R. J. Brachman, L. McGuinness, and L. A. Resnick. 1989. "CLASSIC: A structural data model for objects." In *Proceedings of the ACM SIGMOD International Conference on Management of Data*, pp. 58–67 (New York: ACM Press).

Brachman, R. J., and T. Anand. 1996. "The process of knowledge discovery in databases." In *Advances in Knowledge Discovery and Data Mining*, ed. U. Fayyad, G. Piatetsky-Shapiro, P. Smyth, and R. Uthurusamy, pp. 37–57 (Cambridge, Mass.: MIT Press).

Brachman, R. J., and J. G. Schmolze. 1985. "An overview of the KL-ONE knowledge representation system." *Cognitive Science* 9:171–216.

Branden, C., and J. Tooze. 1991. *Introduction to Protein Structure*. New York: Garland.

Brown, M. T. 1995. "Sequence similarities between the yeast chromosome segregation protein Mif2 and the mammalian centromere protein CENP-C." *Gene* 160:111–116.

Buhmann, J., and H. Kuhnel. 1993. "Complexity of optimized data clustering by competitive neural networks." *Neural Computation* 5:75–88.

Buneman, P., S. B. Davidson, K. Hart, C. Overton, and L. Wong. 1995. "A data transformation system for biological data sources." In *Proceedings of the Twenty-First International Conference on Very Large Data Bases*, pp. 158–169 (San Francisco: Morgan Kaufmann).

Bunke, H., and G. Allermann. 1983. "Inexact graph matching for structural pattern recognition." *Pattern Recognition Letters* 1(4):245–253.

Burks, C. 1989. "GenBank: Current status and future directions." Technical Report LA-UR-89-1154, Los Alamos National Laboratory, Los Alamos, N.M.

Califano, A., and I. Rigoutsos. 1993. "FLASH: A fast lookup algorithm for string homology." In *Proceedings of the First International Conference on Intelligent Systems for Molecular Biology*, pp. 56–64 (Menlo Park, Calif.: AAAI Press).

Carriero, N., and D. Gelernter. 1989. "Linda in context." *Communications of the Association for Computing Machinery* 32(4):444–458.

Cate, J. H., A. R. Gooding, E. Podell, K. Zhou, B. L. Golden, C. E. Kundrot, T. R. Cech, and J. A. Doudna. 1996a. "Crystal structure of a group I ribozyme domain: Principles of RNA parking." *Science* 273:1678–1685.

Cate, J. H., A. R. Gooding, E. Podell, K. Zhou, B. L. Golden, A. A. Szewczak, C. E. Kundrot, T. R. Cech, and J. A. Doudna. 1996b. "RNA tertiary structure mediation by adenosine platforms." *Science* 273:1696–1699.

Chaitin, G. J. 1979. "Toward a mathematical definition of life." In *The Maximum Entropy Formalism*, ed. R. D. Levine and M. Tribus, pp. 477–498 (Cambridge, Mass.: MIT Press).

Chaitin, G. J. 1987. *Algorithmic Information Theory*. New York: Cambridge University Press.

Cheeseman, P., and J. Stutz. 1996. "Bayesian classification (AutoClass): Theory and results." In *Advances in Knowledge Discovery and Data Mining*, ed. U. M. Fayyad, G. Piatetsky-Shapiro, P. Smyth, and R. Uthurusamy, pp. 153–180 (Cambridge, Mass.: MIT Press).

Cheeseman, P., J. Kelly, M. Self, J. Stutz, W. Taylor, and D. Freeman. 1988. "AutoClass: A Bayesian classification system." In *Proceedings of the Fifth International Conference on Machine Learning*, pp. 54–64 (San Francisco: Morgan Kaufmann).

Chen, J., S. Le, B. A. Shapiro, K. M. Currey, and J. V. Maizel. 1990. "A computational procedure for assessing the significance of RNA secondary structure." *Computer Applications in the Biosciences* 6(1):7–18.

Cheng, S. C., E. C. Lynch, K. R. Leason, D. L. Court, B. A. Shapiro, and D. I. Friedman. 1991. "Functional importance of a sequence in the stem-loop of a transcription terminator." *Science* 254:1205–1207.

Chirn, G. W. 1996. *Pattern Discovery in Sequence Databases: Algorithms and Applications to DNA/Protein Classification*. Ph.D. dissertation, Department of Computer and Information Science, New Jersey Institute of Technology.

Chou, P. Y., and G. Fasman. 1978. "Prediction of the secondary structure of proteins from their amino acid sequence." *Advances in Enzymology* 47:145–147.

Christie, B. D., D. R. Henry, W. T. Wipke, and T. E. Moock. 1990. "Database structure and searching in MACCS-3D." *Tetrahedron Computer Methodology* 3(6C):653–664.

Claverie, J.-M., and W. Makalowski. 1994. "Alu alert." *Nature* 371:752.

Claverie, J.-M., and D. States. 1993. "Information enhancement methods for large scale sequence analysis." *Computers in Chemistry* 17:191–201.

Clift, B., D. Haussler, R. McConnell, T. D. Schneider, and G. D. Stormo. 1986. "Sequence landscapes." *Nucleic Acids Research* 14:141–158.

Cobbs, A. L. 1994. "Fast identification of approximately matching substrings." In *Combinatorial Pattern Matching*, ed. M. Crochemore and D. Gusfield, pp. 64–74 (Heidelberg: Springer-Verlag Lecture Notes in Computer Science).

Cochran, W. G. 1977. *Sampling Techniques*. New York: Wiley.

Computer Science and Technology Board. 1990. *Computing and Molecular Biology: Mapping and Interpreting Biological Information, a CSTB Workshop*. Washington, D.C.: National Research Council.

Conklin, D. 1995a. *Knowledge Discovery in Molecular Structure Databases*. Ph.D. dissertation, Department of Computing and Information Science, Queen's University.

Conklin, D. 1995b. "Machine discovery of protein motifs." *Machine Learning* 21:125–150.

Conklin, D., and J. Glasgow. 1992. "Spatial analogy and subsumption." In *Proceedings of the Ninth International Conference on Machine Learning*, pp. 111–116 (San Francisco: Morgan Kaufmann).

Conklin, D., S. Fortier, J. Glasgow, and F. Allen. 1996. "Conformational analysis from crystallographic data using conceptual clustering." *Acta Crystallographica* B52:535–549.

Courteau, J. 1991. "Genome databases." *Science* 254:201–204.

Cover, T., and J. Thomas. 1991. *Elements of Information Theory*. New York: Wiley.

Crippen, G. M., and V. N. Maiorov. 1995. "How many protein folding motifs are there?" *Journal of Molecular Biology* 252:144–151.

Crothers, D. M., and P. E. Cole. 1978. *Transfer RNA*. Cambridge, Mass.: MIT Press.

Currey, K., and B. A. Shapiro. 1997. "Secondary structure computer prediction of the poliovirus 5′ non-coding region is improved with a genetic algorithm." *Computer Applications in the Biosciences* 13:1–12.

Cushing, J. B. 1995. *Computational Proxies: An Object-Based Infrastructure for Experiment Management.* Ph.D. dissertation, Department of Computer Science and Engineering, Oregon Graduate Institute of Science & Technology.

Cushing, J. B., D. Hansen, D. Maier, and C. Pu. 1993. "Connecting scientific programs and data using object databases." *Bulletin of the Technical Committee on Data Engineering* 16(1):9–13.

Cushing, J. B., J. Laird, E. Pasalic, E. Kutter, T. Hunkapiller, F. Zucker, and D. P. Yee. 1997. "Beyond interoperability: Tracking and managing the results of computational applications." In *Proceedings of the Ninth International Conference on Scientific and Statistical Database Management*, pp. 223–236 (Los Alamitos, Calif.: IEEE Computer Society).

Davis, L. 1991. *Handbook of Genetic Algorithms.* New York: Van Nostrand Reinhold.

Dayhoff, M. O., R. M. Schwartz, and B. C. Orcutt. 1978. "A model of evolutionary change in proteins." In *Atlas of Protein Sequence and Structure*, ed. M. O. Dayhoff, Vol. 5, Suppl. 3, pp. 353–358 (Washington, D.C.: National Biomedical Research Foundation).

Dayton, E. T., D. A. M. Konings, D. M. Powell, B. A. Shapiro, L. Butini, J. V. Maizel, Jr., and A. I. Dayton. 1992. "Extensive sequence-specific information throughout the CAR/RRE, the target sequence of the human immunodeficiency virus type 1 Rev protein." *Journal of Virology* 66(2):1139–1151.

Demeler, B., and G. Zhou. 1991. "Neural network optimization for *E. coli* promoter prediction." *Nucleic Acids Research* 19:1593.

Dempster, A. P., N. M. Laird, and D. B. Rubin. 1977. "Maximum likelihood from complete data via the EM algorithm." *Journal of the Royal Statistical Society* 39:1–38.

Derthick, M. 1991. "A minimal encoding approach to feature discovery." In *Proceedings of the Ninth National Conference on Artificial Intelligence*, pp. 565–571 (Menlo Park, Calif.: AAAI Press).

Djoko, S., D. J. Cook, and L. B. Holder. 1996. "Discovering informative structural concepts using domain knowledge." *IEEE Expert* 10:59–68.

Dowe, D., L. Allison, T. Dix, L. Hunter, C. S. Wallace, and T. Edgoose. 1996. "Circular clustering of protein dihedral angles by minimum message length." In *Proceedings of the Pacific Symposium on Biocomputing*, pp. 242–255 (Singapore: World Scientific).

Drmanac, R., I. Labat, I. Brukner, and R. Crkvenjakov. 1989. "Sequencing of megabase plus DNA by hybridization: Theory of the method." *Genomics* 4:114–128.

Drmanac, R., S. Drmanac, Z. Strezoska, T. Paunesku, I. Labat, M. Zeremski, J. Snoddy, W. K. Funkhouser, B. Koop, L. Hood, and R. Crkvenjakov. 1993. "DNA sequence determination by hybridization: A strategy for efficient large-scale sequencing." *Science* 260:1649–1652.

Drmanac, S., and R. Drmanac. 1994. "Processing of cDNA and genomic kilobase-sized clones for massive screening, mapping and sequencing by hybridization." *BioTechniques* 7:328–336.

Dubois, J., G. Carrier, and A. Panaye. 1991. "DARC topological descriptors for pattern recognition in molecular database management systems and design." *Journal of Chemical Information and Computer Sciences* 31(4):574–578.

Fayyad, U. M., G. Piatetsky-Shapiro, P. Smyth, and R. Uthurusamy, eds. 1996a. *Advances in Knowledge Discovery and Data Mining.* Cambridge, Mass.: MIT Press.

Fayyad, U. M., G. Piatetsky-Shapiro, and P. Smyth. 1996b. "From data mining to knowledge discovery: An overview." In *Advances in Knowledge Discovery and Data Mining,* ed. U. M. Fayyad, G. Piatetsky-Shapiro, P. Smyth, and R. Uthurusamy, pp. 1–34 (Cambridge, Mass.: MIT Press).

Feng, D.-F., and R. F. Doolittle. 1987. "Progressive sequence alignment as a prerequisite to correct phylogenetic trees." *Journal of Molecular Evolution* 25:351–360.

Fetrow, J. S., M. J. Palumbo, and G. Berg. 1997. "Patterns, structures, and amino acid frequencies in structural building blocks. A protein secondary structure classification scheme." *PROTEINS: Structure, Function, and Genetics* 27:249–271.

Fickett, J. W., and C. Burks. 1989. *Development of a Database for Nucleotide Sequences.* Boca Raton, Fla.: CRC Press.

Fisanick, W., K. P. Cross, and A. Rusinko, III. 1992. "Similarity searching on CAS registry substances. 1. Global molecular property and generic atom triangle geometric searching." *Journal of Chemical Information and Computer Sciences* 32(6):664–674.

Fisanick, W., A. H. Lipkus, and A. Rusinko, III. 1994. "Similarity searching on CAS registry substances. 2. 2D structural similarity." *Journal of Chemical Information and Computer Sciences* 34(1):130–140.

Fischer, D., O. Bachar, R. Nussinov, and H. J. Wolfson. 1992a. "An efficient automated computer vision based technique for detection of three dimensional structural motifs in proteins." *Journal of Biomolecular Structure & Dynamics* 9(4):769–789.

Fischer, D., R. Nussinov, and H. J. Wolfson. 1992b. "3D substructure matching in protein molecules." In *Combinatorial Pattern Matching,* ed. A. Apostolico, M. Crochemore, Z. Galil, and U. Manber, pp. 136–150 (Heidelberg: Springer-Verlag Lecture Notes in Computer Science).

Fisher, D. 1987. "Knowledge acquisition via incremental conceptual clustering." *Machine Learning* 2:139–172.

Fortier, S., I. Castleden, J. Glasgow, D. Conklin, C. Walmsley, L. Leherte, and F. Allen. 1993. "Molecular scene analysis: The integration of direct methods and artificial intelligence strategies for solving protein crystal structures." *Acta Crystallographica* D49:168–178.

Freier, S., R. Kierzek, J. A. Jaeger, N. Sugimoto, M. Caruthers, T. Neilson, and D. Turner. 1986. "Improved free-energy parameters for predictions of RNA duplex stability." *Proceedings of the National Academy of Sciences of the USA* 83:9373–9377.

Frenkel, K. A. 1991. "The human genome project and informatics." *Communications of the Association for Computing Machinery* 34(11):41–51.

Friezner-Degen, S. J., B. Rajput, and E. Reich. 1986. "The human tissue plasminogen activator gene." *Journal of Biological Chemistry* 261:6972–6985.

Galas, D. J., M. Eggert, and M. S. Waterman. 1985. "Rigorous pattern-recognition methods for DNA sequences: Analysis of promoter sequences from *Escherichia coli.*" *Journal of Molecular Biology* 186:117–128.

Garey, M. R., and D. S. Johnson. 1979. *Computers and Intractability: A Guide to the Theory of NP-Completeness.* New York: W. H. Freeman and Company.

Gautheret, D., S. H. Damberger, and R. R. Gutell. 1995. "Identification of base-triples in RNA using comparative sequence analysis." *Journal of Molecular Biology* 248:27–43.

Geist, A., A. Beguelin, J. Dongarra, W. Jiang, R. Manchek, and V. Sunderam. 1994. *PVM: Parallel Virtual Machine, A User's Guide and Tutorial for Networked Parallel Computing.* Cambridge, Mass.: MIT Press.

Gelfand, M. S. 1995. "Prediction of function in DNA sequence analysis." *Journal of Computational Biology* 2:87–115.

Gennari, J., P. Langley, and D. Fisher. 1989. "Models of incremental concept formation." *Artificial Intelligence* 40:11–61.

Ghosh, D., C. M. Weeks, P. Groschulski, W. L. Duax, M. Erman, R. L. Rimsay, and J. C. Orr. 1991. "Three-dimensional structure of holo 3α-20β-hydroxysteroid dehydrogenase: A member of a short-chain dehydrogenase family." *Proceedings of the National Academy of Sciences of the USA* 88:10064–10068.

Ghosh, D., Z. Wawrzak, C. M. Weeks, W. L. Duax, and M. Erman. 1994a. "The refined three-dimensional structure of 3α-20β-hydroxysteroid dehydrogenase and possible roles of the residues conserved in short-chain dehydrogenases." *Structure* 2:629–640.

Ghosh, D., V. Z. Pletnev, D. W. Zhu, Z. Wawrzak, W. L. Duax, W. Pangborn, F. Labrie, and S. X. Lin. 1994b. "Structure of human estrogenic 17β-hydroxysteroid dehydrogenase at 2.2 Å resolution." *Structure* 3:503–513.

Glasgow, J., S. Fortier, and F. Allen. 1993. "Molecular scene analysis: Crystal structure determination through imagery." In *Artificial Intelligence and Molecular Biology*, ed. L. Hunter, pp. 433–458 (Menlo Park, Calif.: AAAI Press).

Goad, W. B., and M. I. Kanehisa. 1982. "Pattern recognition in nucleic acid sequences. I. A general method for finding local homologies and symmetries." *Nucleic Acids Research* 10:247–263.

Goldberg, D. E., and K. Deb. 1991. "A comparative analysis of selection schemes used in genetic algorithms." In *Foundations of Genetic Algorithms*, ed. G. J. Rawlins, pp. 69–93 (San Francisco: Morgan Kaufmann).

Gonnet, G. H., M. A. Cohen, and S. A. Benner. 1992. "Exhaustive matching of entire protein sequence database." *Science* 256:1443–1445.

Green, P., D. Lipman, L. Hillier, R. Waterston, D. States, and J. M. Claverie. 1993. "Ancient conserved regions in new gene sequences and the protein databases." *Science* 259:1711–1716.

Greer, J. 1991. "Comparative modeling of homologous proteins." *Methods in Enzymology* 202:239–252.

Gribskov, M., A. D. McLachlan, and D. Eisenberg. 1987. "Profile analysis: Detection of distantly related proteins." *Proceedings of the National Academy of Sciences of the USA* 84:4355–4358.

Grundy, W. N., and C. P. Elkan. 1997. "Motif-based hidden Markov models for multiple sequence alignment." Poster presented at the Fifth International Conference on Intelligent Systems for Molecular Biology, Halkidiki, Greece.

Grundy, W. N., T. L. Bailey, C. P. Elkan, and M. E. Baker. 1997. "Hidden Markov model analysis of motifs in steroid dehydrogenases and their homologs." *Biochemistry and Biophysical Research Communications* 231:760–766.

Guan, X., and E. C. Uberbacher. 1996. "A fast lookup algorithm for detecting repetitive DNA sequences." In *Proceedings of the Pacific Symposium on Biocomputing* (Singapore: World Scientific).

Gultyaev, A. P., F. H. D. van Batenburg, and C. W. A. Pleij. 1995. "The computer simulation of RNA folding pathways using a genetic algorithm." *Journal of Molecular Biology* 250:37–51.

Güner, O. F., D. R. Henry, T. E. Moock, and R. S. Pearlman. 1990. "Flexible queries in 3D searching. 2. Techniques in 3D query formulation." *Tetrahedron Computer Methodology* 3(6C):557–563.

Han, K. F., and D. Baker. 1996. "Global properties of the mapping between local amino acid sequence and local structure in proteins." *Proceedings of the National Academy of Sciences of the USA* 93:5814–5818.

Haraki, K. S., R. P. Sheridan, R. Venkataraghavan, D. A. Dunn, and R. McCulloch. 1990. "Looking for pharmacophores in 3D databases: Does conformational searching improve the yield of actives?" *Tetrahedron Computer Methodology* 3(6C):565–573.

Henikoff, J. G., and S. Henikoff. 1996a. "Blocks database and its applications." *Methods in Enzymology* 266:88–105.

Henikoff, J. G., and S. Henikoff. 1996b. "Using substitution probabilities to improve position-specific scoring matrices." *Computer Applications in the Biosciences* 12:135–143.

Henikoff, S. 1991. "Playing with blocks: Some pitfalls of forcing multiple alignments." *New Biologist* 3:1148–1154.

Henikoff, S. 1992. "Detection of *Caenorhabditis* transposon homologs in diverse organisms." *New Biologist* 4:382–388.

Henikoff, S. 1993. "Transcriptional activator components and poxvirus DNA-dependent ATPases comprise a single family." *Trends in Biochemical Sciences* 18:291.

Henikoff, S., and J. G. Henikoff. 1991. "Automated assembly of protein blocks for database searching." *Nucleic Acids Research* 19:6565–6572.

Henikoff, S., and J. G. Henikoff. 1992. "Amino acid substitution matrices from protein blocks." *Proceedings of the National Academy of Sciences of the USA* 89:10915–10919.

Henikoff, S., and J. G. Henikoff. 1994a. "Position-based sequence weights." *Journal of Molecular Biology* 243:574–578.

Henikoff, S., and J. G. Henikoff. 1994b. "Protein family classification based on searching a database of blocks." *Genomics* 19:97–107.

Henikoff, S., and J. G. Henikoff. 1997. "Embedding strategies for effective use of information from multiple sequence alignments." *Protein Science* 6:698–705.

Henikoff, S., J. G. Henikoff, W. G. Alford, and S. Pietrokovski. 1995. "Automated construction and graphical presentation of protein blocks from unaligned sequences." *Gene* 163:GC17–GC26.

Higgins, D. G., J. D. Thompson, and T. J. Gibson. 1996. "Using CLUSTAL for multiple sequence alignments." *Methods in Enzymology* 266:383–402.

Hirst, J. D., and M. J. E. Sternberg. 1992. "Prediction of structural and functional features of protein and nucleic acid sequences by artificial neural networks." *Biochemistry* 31:7211–7219.

Holland, J. H. 1975. *Adaptation in Natural and Artificial Systems.* Ann Arbor: University of Michigan Press.

Hui, L. C. K. 1992. "Color set size problem with applications to string matching." In *Combinatorial Pattern Matching*, ed. A. Apostolico, M. Crochemore, Z. Galil, and U. Manber, pp. 230–243 (Heidelberg: Springer-Verlag Lecture Notes in Computer Science).

Hunkapiller, T., R. J. Kaiser, B. F. Koop, and L. Hood. 1991. "Large-scale and automated DNA sequence determination." *Science* 254:59–67.

Hunter, L., ed. 1993. *Artificial Intelligence and Molecular Biology.* Menlo Park, Calif.: AAAI Press.

Hunter, L., and D. States. 1991. "Applying Bayesian classification to protein structure." In *Proceedings of the Seventh Conference on Artificial Intelligence Applications*, Vol. 1, pp. 10–16 (Miami, Fla.: IEEE Computer Society).

Hurst, T. 1994. "Flexible 3D searching: The directed tweak technique." *Journal of Chemical Information and Computer Sciences* 34(1):190–196.

Ioannidis, Y., M. Livny, S. Gupta, and N. Ponnekanti. 1996. "ZOO: A desktop experiment management environment." In *Proceedings of the Twenty-Second International Conference on Very Large Data Bases*, pp. 274–285 (San Francisco: Morgan Kaufmann).

Jacobson, S. J., D. A. M. Konings, and P. Sarnow. 1993. "Biochemical and genetic evidence for a pseudoknot structure at the 3' terminus of the poliovirus RNA genome and its role in viral RNA amplification." *Journal of Virology* 67:2961–2971.

Jaeger, J. A., D. H. Turner, and M. Zuker. 1989a. "Improved predictions of secondary structures for RNA." *Proceedings of the National Academy of Sciences of the USA* 86:7706–7710.

Jaeger, J. A., D. Turner, and M. Zuker. 1989b. "Predicting optimal and suboptimal secondary structure for RNA." *Methods in Enzymology* 183:281–306.

Jarvis, R. A., and E. A. Patrick. 1973. "Clustering using a similarity measure based on shared near neighbors." *IEEE Transactions on Computers* C-22(11):1025–1034.

Jeong, K. 1996. *Fault-Tolerant Parallel Processing Combining Linda, Checkpointing, and Transactions.* Ph.D. dissertation, Computer Science Department, Courant Institute of Mathematical Sciences, New York University.

Johnson, M. S., and J. P. Overington. 1993. "A structural basis for sequence comparisons: An evaluation of scoring methodologies." *Journal of Molecular Biology* 233:716–738.

Jones, D., W. Taylor, and J. Thornton. 1992. "A new approach to protein fold recognition." *Nature* 358:86–89.

Jones, T., and S. Thirup. 1986. "Using known substructures in protein model building and crystallography." *The European Molecular Biology Organization Journal* 5(4):819–822.

Jornvall, H., B. Persson, M. Krook, S. Atrian, R. Gonzalez-Duarte, J. Jeffrey, and D. Ghosh. 1995. "Short-chain dehydrogenases/reductases (SDR)." *Biochemistry* 34:6003–6013.

Jurka, J., and A. Milosavljević. 1991. "Reconstruction and analysis of human Alu genes." *Journal of Molecular Evolution* 32:105–121.

Jurka, J., D. J. Kaplan, C. H. Duncan, J. Walichiewicz, A. Milosavljević, G. Murali, and J. F. Solus. 1993. "Identification and characterization of new human medium reiteration frequency repeats." *Nucleic Acids Research* 21:1273–1279.

Kabsch, W. 1976. "A solution for the best rotation to relate two sets of vectors." *Acta Crystallographica* A32:922–923.

Kabsch, W. 1978. "A discussion of the solution for the best rotation to relate two sets of vectors." *Acta Crystallographica* A34:827–828.

Kabsch, W., and C. Sander. 1983. "Dictionary of protein secondary structure: Pattern recognition of hydrogen-bonded and geometrical features." *Biopolymers* 22:2577–2637.

Karp, P. D., M. Riley, S. M. Paley, and A. Pelligrini-Toole. 1996. "EcoCyc: An encyclopedia of *Escherichia coli* genes and metabolism." *Nucleic Acids Research* 24:32–39.

Karpen, M. E., P. L. de Haseth, and K. E. Neet. 1989. "Comparing short protein substructures by a method based on backbone torsion angles." *PROTEINS: Structure, Function, and Genetics* 6:155–167.

Karypis, G., and V. Kumar. 1995. "Multilevel *k*-way partitioning scheme for irregular graphs." Technical Report, Department of Computer Science, University of Minnesota.

Kemp, G. J. L., J. Dupont, and P. M. D. Gray. 1996. "Using the functional data model to integrate distributed biological data sources." In *Proceedings of the Eighth International Conference on Scientific and Statistical Database Management*, pp. 176–185 (Los Alamitos, Calif.: IEEE Computer Society).

Kim, S. H., F. L. Suddah, G. J. Quigley, A. McPherson, J. L. Sussman, A. H. J. Wang, N. C. Seeman, and A. Rich. 1974. "Three-dimensional tertiary structure of yeast phenylalanine transfer RNA." *Science* 185:435–440.

Kiviat, P. J., H. M. Markowitz, and R. Villanueva. 1983. *SIMSCRIPT II.5 Programming Language.* Los Angeles: CACI.

Klingler, T. M., and D. L. Brutlag. 1994. "Discovering structural correlations in α-helices." *Protein Science* 3:1847–1857.

Kneller, D. G., F. E. Cohen, and R. Langridge. 1990. "Improvements in protein secondary structure prediction by an enhanced neural network." *Journal of Molecular Biology* 214:171–182.

Koch, I., T. Lengauer, and E. Wanke. 1996. "An algorithm for finding maximal common subtopologies in a set of protein structures." *Journal of Computational Biology* 3:289–306.

Kolaskar, A. S., and U. Kulkarni-Kale. 1992. "Sequence alignment approach to pick up conformationally similar protein segments." *Journal of Molecular Biology* 223:1053–1061.

Korber, B., R. Farber, D. Wolpert, and A. Lapedes. 1993. "Covariation of mutations in the V3 loop of human immunodeficiency virus type 1 envelope protein: An information-theoretic analysis." *Proceedings of the National Academy of Sciences of the USA* 90(15):7176–7180.

Krnjajic, M. 1996. "Three methods for finding repetitive DNA patterns." CMP243 Course Project, Computer Science Department, University of California at Santa Cruz.

Krogh, A., M. Brown, I. S. Mian, K. Sjolander, and D. Haussler. 1994. "Hidden Markov models in computational biology: Applications to protein modeling." *Journal of Molecular Biology* 235:1501–1531.

Krozowski, Z. 1992. "11β-hydroxysteroid dehydrogenase and the short chain alcohol dehydrogenase (SCAD) superfamily." *Molecular and Cellular Endocrinology* 84:C25–C31.

Kuntz, I. D. 1992. "Structure-based strategies for drug design and discovery." *Science* 257:1078–1082.

Laiter, S., D. L. Hoffman, R. K. Singh, I. I. Vaisman, and A. Tropsha. 1995. "Pseudotorsional OCCO backbone angle as a single descriptor of protein secondary structure." *Protein Science* 4:1633–1643.

Lamdan, Y., J. T. Schwartz, and H. J. Wolfson. 1990. "Affine-invariant model-based object recognition." *IEEE Transactions on Robotics and Automation* 6(5):577–589.

Landau, G. M., and U. Vishkin. 1989. "Fast parallel and serial approximate string matching." *Journal of Algorithms* 10:157–169.

Lander, E. S., and P. Green. 1987. "Construction of multilocus genetic linkage maps in humans." *Proceedings of the National Academy of Sciences of the USA* 84:2363–2367.

Lander, E. S., and M. S. Waterman, eds. 1995. *Calculating the Secrets of Life.* Washington, D.C.: National Academy Press.

Lander, E. S., R. Langridge, and D. Saccocio. 1991. "Computing in molecular biology: Mapping and interpreting biological information." *IEEE Computer* 24(11):6–13.

Lapedes, A., C. Barnes, C. Burks, R. Farber, and K. Sirotkin. 1990. "Application of neural networks and other machine learning algorithms to DNA sequence analysis." In *Computers and DNA*, ed. G. I. Bell and T. G. Marr, pp. 157–182 (Reading, Mass.: Addison-Wesley).

Lapedes, A., E. Steeg, and R. Farber. 1995. "Use of adaptive networks to evolve highly predictable protein secondary-structure classes." *Machine Learning* 21:103–124.

Lathrop, R. H., T. A. Webster, R. F. Smith, P. H. Winston, and T. F. Smith. 1993. "Integrating AI with sequence analysis." In *Artificial Intelligence and Molecular Biology*, ed. L. Hunter, pp. 210–258 (Menlo Park, Calif.: AAAI Press).

Lawrence, C., and A. Reilly. 1990. "An expectation maximization (EM) algorithm for the identification and characterization of common sites in unaligned biopolymer sequences." *Proteins* 7:41–51.

Lawrence, C. E., S. F. Altschul, M. S. Boguski, J. S. Liu, A. F. Neuwald, and J. C. Wootton. 1993. "Detecting subtle sequence signals: A Gibbs sampling strategy for multiple alignment." *Science* 262:208–214.

Le, S., and M. Zuker. 1990. "Common structures of the 5' noncoding RNA in enteroviruses and rhinoviruses: Thermodynamical stability and statistical significance." *Journal of Molecular Biology* 216:729–741.

Le, S., J. Chen, K. Currey, and J. V. Maizel. 1988. "A program for predicting significant RNA secondary structures." *Computer Applications in the Biosciences* 4:153.

Le, S., J. Owens, R. Nussinov, J.-H. Chen, B. A. Shapiro, and J. V. Maizel. 1989. "RNA secondary structures: Comparison and determination of frequently recurring substructures by consensus." *Computer Applications in the Biosciences* 5(3):205–210.

Le, S., B. A. Shapiro, J. Chen, R. Nussinov, and J. V. Maizel. 1991. "RNA pseudoknots downstream of the frameshift sites of retroviruses." *Genetic Analysis Techniques and Applications* 8(7):191–205.

Lebowitz, M. 1987. "Experiments with incremental concept formation: UNIMEM." *Machine Learning* 2:103–138.

Leclerc, Y. G. 1989. "Constructing simple stable descriptions for image partitioning." *International Journal of Computer Vision* 3(1):73–102.

Letovsky, S., R. Pecherer, and A. Shoshani. 1990. "Scientific data management for human genome applications." *Bulletin of the Technical Committee on Data Engineering* 13(3):51.

Li, M., and P. M. B. Vitányi. 1993. *An Introduction to Kolmogorov Complexity and Its Applications.* New York: Springer-Verlag.

Lipman, D. J., and W. R. Pearson. 1985. "Rapid and sensitive protein similarity searches." *Science* 227:1435–1441.

Lipman, D. J., S. F. Altschul, and J. D. Kececioglu. 1989. "A tool for multiple sequence alignment." *Proceedings of the National Academy of Sciences of the USA* 86:4412–4415.

Loewenstern, D., H. Hirsh, P. Yianilos, and M. Noordewier. 1995. "DNA sequence classification using compression-based induction." DIMACS Technical Report 95-04, Rutgers University.

Losee, J. 1980. *A Historical Introduction to the Philosophy of Science.* New York: Oxford University Press.

Lukashin, A. V., V. V. Anshelevich, B. R. Amirikyan, A. I. Gragerov, and M. D. Frank-Kamenetskii. 1989. "Neural network models for promoter recognition." *Journal of Biomolecular Structure & Dynamics* 6:1123–1133.

Lüthy, R., A. D. McLachlan, and D. Eisenberg. 1991. "Secondary structure-based profiles: Use of structure-conserving scoring tables in searching protein sequence databases for structural similarities." *PROTEINS: Structure, Function, and Genetics* 10:229–239.

MacKay, D. J. C. 1992. *Bayesian Methods for Adaptive Models.* Ph.D. dissertation, California Institute of Technology.

Mannila, H. 1996. "Data mining: Machine learning, statistics, and databases." In *Proceedings of the Eighth International Conference on Scientific and Statistical Database Management*, pp. 2–8 (Los Alamitos, Calif.: IEEE Computer Society).

Manola, F., and U. Dayal. 1990. "PDM: An object-oriented data model." In *Readings in Object-Oriented Database Systems*, ed. S. B. Zdonik and D. Maier, pp. 209–215 (San Francisco: Morgan Kaufmann).

Margalit, H., B. A. Shapiro, A. Oppenheim, and J. Maizel. 1989. "Detection of common motifs in RNA secondary structure." *Nucleic Acids Research* 17:4829–4845.

Marr, T. G. 1996. *Genome Topographer User Manual.* Boulder, Colo.: Genomica Corporation.

Martin, Y., M. G. Bures, and P. Willett. 1990. "Searching databases of three-dimensional structures." In *Reviews In Computational Chemistry*, ed. K. Lipkowitz and D. Boyd, pp. 213–263 (New York: VCH).

Martinez, H. M. 1988. "An RNA secondary structure workbench." *Nucleic Acids Research* 16:1789–1798.

Martinez, H. M. 1990. "Detecting pseudoknots and other local base-pairing structures in RNA structure prediction." *Methods in Enzymology* 183:306–317.

Matsuo, Y., and M. Kanehisa. 1993. "An approach to systematic detection of protein structural motifs." *Computer Applications in the Biosciences* 9:153–159.

McCreight, E. M. 1976. "A space-economical suffix tree construction algorithm." *Journal of the Association for Computing Machinery* 23:262–272.

McGregor, M., T. Flores, and M. Sternberg. 1989. "Prediction of β-turns in proteins using neural networks." *Protein Engineering* 2:521–526.

McLachlan, G., and K. Basford. 1988. *Mixture Models: Inference and Applications to Clustering.* New York: Marcel Dekker.

Metropolis, N., A. W. Rosenbluth, M. N. Rosenbluth, A. H. Teller, and E. Teller. 1953. "Equation of state calculations by fast computing machines." *Journal of Chemical Physics* 21:1087–1092.

Milne, G. W. A., M. C. Nicklaus, J. S. Driscoll, S. Wang, and D. Zaharevitz. 1994. "National Cancer Institute drug information system 3D database." *Journal of Chemical Information and Computer Sciences* 34(5):1219–1224.

Milosavljević, A. 1995a. "Discovering dependencies via algorithmic mutual information: A case study in DNA sequence comparisons." *Machine Learning* 21:35–50.

Milosavljević, A. 1995b. "The discovery process as a search for concise encoding of observed data." *Foundations of Science* 1(2):212–218.

Milosavljević, A. 1995c. "DNA sequence recognition by hybridization to short oligomers." *Journal of Computational Biology* 2(2):355–370.

Milosavljević, A., and J. Jurka. 1993a. "Discovering simple DNA sequences by the algorithmic significance method." *Computer Applications in the Biosciences* 9(4):407–411.

Milosavljević, A., and J. Jurka. 1993b. "Discovery by minimal length encoding: A case study in molecular evolution." *Machine Learning, Special Issue on Machine Discovery* 12(1–3):69–87.

Milosavljević, A., M. Zeremski, Ž. Strezoska, D. Grujić, H. Dyanov, A. Gemmell, S. Batus, D. Salbego, T. Paunesku, B. Soares, and R. Crkvenjakov. 1996a. "Discovering distinct genes represented in 29, 570 clones from infant brain cDNA libraries by applying sequencing by hybridization methodology." *Genome Research* 6:132–141.

Milosavljević, A., S. Savković, R. Crkvenjakov, D. Salbego, H. Serrato, H. Kreuzer, A. Gemmell, S. Batus, D. Grujic, S. Carnahan, T. Paunesku, and J. Tepavčević. 1996b. "DNA sequence recognition by hybridization to short oligomers: Experimental verification of the method on the *E. coli* genome." *Genomics* 37(1):77–86.

Mironov, A. A., L. P. Dyakonova, and A. E. Kister. 1985. "A kinetic approach to the prediction of RNA secondary structures." *Journal of Biomolecular Structure & Dynamics* 2:953–962.

Moon, J. B., and W. J. Howe. 1990. "3D database searching and *de novo* construction methods in molecular design." *Tetrahedron Computer Methodology* 3(6C):697–711.

Mulligan, M. E., and W. R. McClure. 1986. "Analysis of the occurrence of promoter-sites in DNA." *Nucleic Acids Research* 14:109–126.

Murrall, N. W., and E. K. Davies. 1990. "Conformational freedom in 3D databases. 1. Techniques." *Journal of Chemical Information and Computer Sciences* 30:312–316.

Needleman, S. B., and C. D. Wunsch. 1970. "A general method applicable to the search for similarities in the amino acid sequence of two proteins." *Journal of Molecular Biology* 48:443–453.

Neuwald, A. F., and P. Green. 1994. "Detecting patterns in protein sequences." *Journal of Molecular Biology* 239:698–712.

Neuwald, A. F., J. S. Liu, and C. E. Lawrence. 1995. "Gibbs motif sampling: Detection of bacterial outer membrane protein repeats." *Protein Science* 4:1618–1632.

Niefind, K., and D. Schomburg. 1991. "Amino acid similarity coefficients for protein modeling and sequence alignment derived from main-chain folding angles." *Journal of Molecular Biology* 219:481–497.

Nilakantan, R., N. Bauman, and R. Venkataraghavan. 1993. "New method for rapid characterization of molecular shapes: Applications in drug design." *Journal of Chemical Information and Computer Sciences* 33(1):79–85.

Nussinov, R., and H. J. Wolfson. 1991. "Efficient detection of three-dimensional motifs in biological macromolecules by computer vision techniques." *Proceedings of the National Academy of Sciences of the USA* 88:10495–10499.

Ogiwara, A., I. Uchiyama, Y. Seto, and M. Kanehisa. 1992. "Construction of a dictionary of sequence motifs that characterize groups of related proteins." *Protein Engineering* 5:479–488.

Ogiwara, A., I. Uchiyama, T. Takagi, and M. Kanehisa. 1996. "Construction and analysis of a profile library characterizing groups of structurally known proteins." *Protein Science* 5:1991–1999.

Oldfield, T. J., and R. E. Hubbard. 1994. "Analysis of C_α geometry in protein structures." *PROTEINS: Structure, Function, and Genetics* 18:324–337.

OLE Team. 1996. *What OLE Is Really About.* Seattle: Microsoft Corporation.

O'Neill, M. C., and F. Chiafari. 1989. "*Escherichia coli* promoters. II. A spacing class-dependent promoter search protocol." *Journal of Biological Chemistry* 264:5531–5534.

Orengo, C. A. 1992. "A review of methods for protein structure comparison." In *Patterns in Protein Sequence and Structure, Biophysics,* Vol. 7, ed. W. R. Taylor, pp. 159–188 (Heidelberg: Springer-Verlag).

Orengo, C. A., D. T. Jones, and J. Thornton. 1994. "Protein superfamilies and domain superfolds." *Nature* 372:631–634.

Overington, J., D. Donnely, M. S. Johnson, A. Sali, and T. L. Blundell. 1992. "Environment-specific amino acid substitution tables: Tertiary templates and prediction of protein folds." *Protein Science* 1:216–226.

ParcPlace. 1997. "Distributed Smalltalk." Available at http://www.parcplace.com/products/dst/info/dst.htm.

Parry-Smith, D. J., and T. K. Attwood. 1992. "ADSP—a new package for computational sequence analysis." *Computer Applications in the Biosciences* 8:451–459.

Pauling, L., R. Corey, and H. Branson. 1951. "The structure of proteins: Two hydrogen-bonded helical configurations of the polypeptide chain." *Proceedings of the National Academy of Sciences of the USA* 37:205–211.

Pearson, W. R., and D. J. Lipman. 1988. "Improved tools for biological sequence comparison." *Proceedings of the National Academy of Sciences of the USA* 85:2444–2448.

Pentland, A. 1989. "Part segmentation for object recognition." *Neural Computation* 1:82–91.

Pepperrell, C. A., R. Taylor, and P. Willett. 1990. "Implementation and use of an atom-mapping procedure for similarity searching in databases of 3D chemical structures." *Tetrahedron Computer Methodology* 3(6C):575–593.

Persson, B., M. Krook, and H. Jornvall. 1991. "Characteristics of short-chain alcohol dehydrogenases and related enzymes." *European Journal of Biochemistry* 200:537–543.

Peskin, R. L., and S. S. Walther. 1994. "SCENE—a computational interface system." *Computers in Physics* 8:430–437.

Pietrokovski, S. 1994. "Conserved sequence features of inteins (proton introns) and their use in identifying new inteins and related proteins." *Protein Science* 3:2340–2350.

Pleij, C. W. A. 1990. "Pseudoknots: a new motif in the RNA game." *Trends in Biochemical Sciences* 15:143–147.

Pleij, C. W. A., K. Rietveld, and L. Bosch. 1985. "A new principle of RNA folding based on pseudoknotting." *Nucleic Acids Research* 13:1717–1731.

Pley, H. A., K. M. Flaherty, and D. B. McKay. 1994. "Three-dimensional structure of a hammerhead ribozyme." *Nature* 372:68–74.

Ponder, J., and F. Richards. 1987. "Tertiary templates for proteins—use of packing criteria in the enumeration of allowed sequences for different structural classes." *Journal of Molecular Biology* 193:775–791.

Posfai, J., A. S. Bhagwat, and R. J. Roberts. 1988. "Sequence motifs specific for cytosine methyltransferases." *Gene* 74:261–265.

Press, W. H., S. A. Teukolsky, W. T. Vetterling, and B. P. Flannery. 1992. *Numerical Recipes in FORTRAN: The Art of Scientific Computing*, 2nd ed., pp. 436–438. New York: Cambridge University Press.

Prestrelski, S. J., A. L. Williams, Jr., and M. N. Liebman. 1992. "Generation of a substructure library for the description and classification of protein secondary structure. I. Overview of the methods and results." *PROTEINS: Structure, Function, and Genetics* 14:430–439.

Pu, C., K. P. Sheka, L. Chang, J. Ong, A. Chang, E. Alessio, I. N. Shindyalov, W. Chang, and P. E. Bourne. 1992. "PDBtool: A prototype object oriented toolkit for protein structure verification." Technical Report CUCS-048-92, Department of Computer Science, Columbia University.

Qian, N., and T. J. Sejnowski. 1988. "Predicting the secondary structure of globular proteins using neural network models." *Journal of Molecular Biology* 202:865–884.

Qu, C., L. Lai, X. Xu, and Y. Tang. 1993. "Phyletic relationships of protein structures based on spatial preference of residues." *Journal of Molecular Evolution* 36:67–78.

Quinlan, J. R. 1986. "Induction of decision trees." *Machine Learning* 1:81–106.

Quinlan, J. R., and R. L. Rivest. 1989. "Inferring decision trees using the minimum description length principle." *Information and Computation* 80:227–248.

Quinqueton, J., and J. Moreau. 1985. "Application of learning techniques to splicing site recognition." *Biochimie* 67:541–548.

Rackovsky, S. 1990. "Quantitative organization of the known protein X-ray structures. I. Methods and short-length-scale results." *PROTEINS: Structure, Function, and Genetics* 7:378–402.

Rafferty, J. B., J. W. Simon, C. Baldock, P. J. Artymiuk, P. J. Baker, A. R. Stuitje, A. R. Slabas, and D. W. Rice. 1995. "Common themes in redox chemistry emerge from the X-ray structure of oilseed rape (*Brassica napus*) enoyl acyl carrier protein reductase." *Structure* 3:927–938.

Rao, R. B., and S. C. Lu. 1992. "Learning engineering models with the minimum description length principle." In *Proceedings of the Tenth National Conference on Artificial Intelligence*, pp. 717–722 (Menlo Park, Calif.: AAAI Press).

Rieche, B., and K. R. Dittrich. 1994. "A federated DBMS-based integrated environment for molecular biology." In *Proceedings of the Seventh International Working Conference on Scientific and Statistical Database Management*, pp. 118–127 (Los Alamitos, Calif.: IEEE Computer Society).

Riesner, D., and R. Romer. 1973. *Physiochemical Properties of Nucleic Acids*. New York: Academic Press.

Rigoutsos, I. 1992. *Massively Parallel Bayesian Object Recognition*. Ph.D. dissertation, Computer Science Department, Courant Institute of Mathematical Sciences, New York University.

Rigoutsos, I. 1997. "Affine-invariants that distribute uniformly and can be tuned to any convex feature domain. Case II: Three-dimensional feature domains." Technical Report RC 20879, IBM T. J. Watson Research Center, New York.

Rigoutsos, I., and A. Delis. 1996. "Managing statistical behavior in very large data sets." Technical Report RC 20648, IBM T. J. Watson Research Center, New York.

Risler, J. L., M. O. Delorme, H. Delacroix, and A. Henaut. 1988. "Amino acid substitutions in structurally related proteins: A pattern recognition approach. Determination of a new and efficient scoring matrix." *Journal of Molecular Biology* 204:1019–1029.

Rissanen, J. 1989. *Stochastic Complexity in Statistical Inquiry*. Singapore: World Scientific.

Rivals, E., O. Delgrande, J.-P. Delahaye, M. Dauchet, M.-O. Delorme, A. Hénaut, and E. Ollivier. 1997. "Detection of significant patterns by compression algorithms: The case of approximate tandem repeats in DNA sequences." *Computer Applications in the Biosciences* 13(2):131–136.

Rodriguez-Tome, P., P. J. Stoehr, G. N. Cameron, and T. P. Flores. 1996. "The European Bioinformatics Institute (EBI) databases." *Nucleic Acids Research* 24:6–12.

Rooman, M. J., and S. J. Wodak. 1988. "Identification of predictive sequence motifs limited by protein structure data base size." *Nature* 335:45–49.

Rooman, M. J., J. Rodriguez, and S. J. Wodak. 1990. "Automatic definition of recurrent local structure motifs in proteins." *Journal of Molecular Biology* 213:327–336.

Roytberg, M. A. 1992. "A search for common patterns in many sequences." *Computer Applications in the Biosciences* 8:57–64.

Sakaguchi, K., N. Zambrano, E. T. Baldwin, B. A. Shapiro, J. W. Erickson, J. G. Omichinski, G. M. Clore, A. M. Gronenborn, and E. Appella. 1993. "Identification of a binding site for the human immunodeficiency virus type 1 nucleocapsid protein." *Proceedings of the National Academy of Sciences of the USA* 90:5219–5223.

Sallantin, J., J. Haiech, and F. Rodier. 1985. "Search for promoter sites of prokaryotic DNA using learning techniques." *Biochimie* 67:549–553.

Sandak, B., R. Nussinov, and H. J. Wolfson. 1995. "An automated computer vision and robotics-based technique for 3D flexible biomolecular docking and matching." *Computer Applications in the Biosciences* 11(1):87–99.

Sankoff, D., and J. B. Kruskal, eds. 1983. *Time Warps, String Edits, and Macromolecules: The Theory and Practice of Sequence Comparison*. Reading, Mass.: Addison-Wesley.

Satou, K., G. Shibayama, T. Ono, Y. Yamamura, E. Furuichi, S. Kuhara, and T. Takagi. 1997. "Finding association rules on heterogeneous genome data." In *Proceedings of the Pacific Symposium on Biocomputing*, pp. 397–408 (Singapore: World Scientific).

Sayle, R. 1994. *RasMol v2.5. A Molecular Visualization Program*. Middlesex, UK: Glaxo Research and Development.

Schuchhardt, J., G. Schneider, J. Reichelt, D. Schomburg, and P. Wrede. 1996. "Local structural motifs of protein backbones are classified by self-organizing neural networks." *Protein Engineering* 9:833–842.

Schuler, G. D., S. F. Altschul, and D. J. Lipman. 1991. "A workbench for multiple alignment construction and analysis." *PROTEINS: Structure, Function, and Genetics* 9:180–190.

Schulze-Kremer, S., and R. King. 1992. "IPSA—inductive protein structure analysis." *Protein Engineering* 5(5):377–390.

Shapiro, B. A. 1988. "An algorithm for comparing multiple RNA secondary structures." *Computer Applications in the Biosciences* 4:378–393.

Shapiro, B. A., and W. Kasprzak. 1996. "Structurelab: A heterogeneous bioinformatics system for RNA structure analysis." *Journal of Molecular Graphics* 14:194–205.

Shapiro, B. A., and J. Navetta. 1994. "A massively parallel genetic algorithm for RNA secondary structure prediction." *Journal of Supercomputing* 8:195–207.

Shapiro, B. A., and J. C. Wu. 1996. "An annealing mutation operator in the genetic algorithms for RNA folding." *Computer Applications in the Biosciences* 12:171–180.

Shapiro, B. A., and J. C. Wu. 1997. "Predicting RNA H-type pseudoknots with the massively parallel genetic algorithm." *Computer Applications in the Biosciences* 13(4):459–471.

Shapiro, B. A., and K. Zhang. 1990. "Comparing multiple RNA secondary structures using tree comparisons." *Computer Applications in the Biosciences* 6(4):309–318.

Shapiro, B. A., L. Lipkin, and J. Maizel. 1982. "An interactive technique for the display of nucleic acid secondary structure." *Nucleic Acids Research* 10:7041–7052.

Shapiro, B. A., J. Maizel, L. Lipkin, K. Currey, and C. Whitney. 1984. "Generating non-overlapping displays of nucleic acid secondary structure." *Nucleic Acids Research* 12:75–88.

Shapiro, B. A., J. Chen, T. Busse, J. Navetta, W. Kasprzak, and J. Maizel. 1995. "Optimization and performance analysis of a massively parallel dynamic programming algorithm for RNA secondary structure prediction." *The International Journal of Supercomputer Applications* 9(1):29–39.

Shimozono, S., A. Shinohara, T. Shinohara, S. Miyano, S. Kuhara, and S. Arikawa. 1992. "Finding alphabet indexing for decision trees over regular patterns: An approach to bioinformatical knowledge acquisition." Technical Report RIFIS-TR-CS-60, Research Institute of Fundamental Information Science, Kyusha University.

Shneiderman, B., ed. 1993. *Sparks of Innovation in Human-Computer Interaction.* Norwood, N.J.: Ablex.

Shpaer, E. G., M. Robinson, D. P. Yee, J. D. Candlin, R. Mines, and T. Hunkapiller. 1996. "Sensitivity and selectivity in protein similarity searches: A comparison of Smith-Waterman in hardware to BLAST and Fasta." *Genomics* 38:179–191.

Sippl, M. J. 1990. "The calculation of conformational ensembles from potentials of mean force. An approach to the knowledge-based prediction of local structures of globular proteins." *Journal of Molecular Biology* 213:859–883.

Skinner, M. A., V. R. Racaniello, G. Dunn, J. Cooper, P. D. Minor, and J. W. Almond. 1989. "New model for the secondary structure of the 5′ noncoding RNA of poliovirus is supported by biochemical and genetic data that also show the RNA secondary structure is important in neurovirulence." *Journal of Molecular Biology* 207:379–392.

Smith, D. W., ed. 1994. *Biocomputing: Informatics and Genome Projects.* San Diego: Academic Press.

Smith, H. O., T. M. Annau, and S. Chandrasegaran. 1990. "Finding sequence motifs in groups of functionally related proteins." *Proceedings of the National Academy of Sciences of the USA* 87:826–830.

Smith, T. F., and M. S. Waterman. 1981a. "Comparison of biosequences." *Advances in Applied Mathematics* 2:482–489.

Smith, T. F., and M. S. Waterman. 1981b. "Identification of common molecular subsequences." *Journal of Molecular Biology* 147:195–197.

Sobel, E., and H. M. Martinez. 1986. "A multiple sequence alignment program." *Nucleic Acids Research* 14:363–374.

Sober, E. 1988. *Reconstructing the Past: Parsimony, Evolution, and Inference.* Cambridge, Mass.: MIT Press.

Solomonoff, R. J. 1964. "A formal theory of inductive inference, Part I." *Information and Control* 7:1–22.

Sparr, T. M., R. D. Bergeron, L. D. Meeker, N. Kinner, P. Mayewski, and M. Person. 1991. "Integrating data management, analysis and visualization for collaborative scientific research." Technical Report 91-10, University of New Hampshire.

Staden, R. 1984. "Computer methods to locate signals in nucleic acid sequences." *Nucleic Acids Research* 12:505–519.

Steeg, E. W. 1997. *Automated Motif Discovery in Protein Structure Prediction.* Ph.D. dissertation, Department of Computer Science, University of Toronto.

Stockman, G. 1987. "Object recognition and localization via pose clustering." *Computer Vision, Graphics, and Image Processing* 40:361–387.

Storer, J. A. 1988. *Data Compression: Methods and Theory.* Rockville, Md.: Computer Science Press.

Studnicka, G. M. 1987. "Nucleotide sequence homologies in control regions of prokaryotic genomes." *Gene* 58:45–57.

SunSoft. 1993. *The ToolTalk Service: An Inter-Operability Solution.* Englewood Cliffs, N.J.: Prentice Hall, SunSoft Press.

Sutton, G., O. White, M. Adams, and A. Kerlavage. 1995. "TIGR assembler: A new tool for assembling large shotgun sequencing projects." *Genome Science & Technology* 1:9–19.

Swindells, M. B. 1993. "Classification of doubly wound nucleotide binding topologies using automated loop searches." *Protein Science* 2:2146–2153.

SYBYL Release 6.01. 1997. Tripos Associates Inc., 1699 S. Hanley Rd., St. Louis, Mo.

Tannin, G. M., A. K. Agrawal, C. Monder, M. I. New, and P. C. White. 1991. "The human gene for 11β-hydroxysteroid dehydrogenase." *Journal of Biological Chemistry* 266:16653–16658.

Tatusov, R. L., S. F. Altschul, and E. V. Koonin. 1994. "Detection of conserved segments in proteins: Iterative scanning of sequence databases with alignment blocks." *Proceedings of the National Academy of Sciences of the USA* 91:12091–12095.

Taylor, W. R. 1986a. "The classification of amino acid conservations." *Journal of Theoretical Biology* 119:205–218.

Taylor, W. R. 1986b. "Identification of protein sequence homology by consensus template." *Journal of Molecular Biology* 188:233–258.

Taylor, W. R., and J. Thornton. 1984. "Recognition of super-secondary structure in proteins." *Journal of Molecular Biology* 173:487–514.

ten Dam, E., K. Pleij, and D. Draper. 1992. "Structural and functional aspects of RNA pseudoknots." *Biochemistry* 31(47):11665–11676.

Thompson, M. J., and R. A. Goldstein. 1996. "Constructing amino acid residue substitution classes maximally indicative of local protein structure." *PROTEINS: Structure, Function, and Genetics* 25:28–37.

Thornton, J., and S. Gardner. 1989. "Protein motifs and data-base searching." *Trends in Biochemical Sciences* 14:300–304.

Tomii, K., and M. Kanehisa. 1996. "Analysis of amino acid indices and mutation matrices for sequence comparison and structure prediction of proteins." *Protein Engineering* 9:27–36.

Tu, Z., N. M. Chapman, G. Hufnagel, S. Tracy, J. R. Romero, W. H. Barry, L. Zhao, K. Currey, and B. A. Shapiro. 1995. "The cardiovirulent phenotype of coxsackievirus b3 is determined at a single site in the genomic 5′ untranslated region." *Journal of Virology* 69:4607–4618.

Tufty, R. M., and R. H. Kretsinger. 1975. "Troponin and parvalbumin calcium binding regions predicted in myosin light chain and T4 lysozyme." *Science* 187:167–169.

Turner, D. H., N. Sugimoto, J. A. Jaeger, C. E. Longfellow, S. M. Freier, and R. Kierzek. 1987. "Improved parameters for predictions of RNA structure." *Cold Spring Harbor Symposia on Quantitative Biology* 52:123–133.

Turner, D. H., N. Sugimoto, and S. M. Freier. 1988. "RNA structure predictions." *Annual Review of Biophysics and Biophysical Chemistry* 17:167–192.

Ullmann, J. R. 1976. "An algorithm for subgraph isomorphism." *Journal of the Association for Computing Machinery* 23(1):31–42.

Unger, R. 1994. "Short structural motifs: definition, identification, and applications." In *The Protein Folding Problem and Tertiary Structure Prediction,* ed. K. Merz, Jr. and S. Le Grand, pp. 339–351 (Boston: Birkhäuser).

Unger, R., and J. L. Sussman. 1993. "The importance of short structural motifs in protein structure analysis." *Journal of Computer-Aided Molecular Design* 7:457–472.

Unger, R., D. Harel, S. Wherland, and J. L. Sussman. 1989. "A 3D building blocks approach to analyzing and predicting structure of proteins." *PROTEINS: Structure, Function, and Genetics* 5:355–373.

Usha, R., and M. R. N. Murthy. 1986. "Protein structural homology: A metric approach." *International Journal of Peptide and Protein Research* 28:364–369.

van Batenburg, F. H. D., A. P. Gultyaev, and C. W. A. Pleij. 1995. "An APL-programmed genetic algorithm for the prediction of RNA secondary structure." *Journal of Theoretical Biology* 174:269–280.

Varughese, K. I., M. M. Skinner, J. M. Whitely, D. A. Matthews, and N. H. Xuong. 1992. "Crystal structure of rat liver dihydropteridine reductase." *Proceedings of the National Academy of Sciences of the USA* 89:6080–6084.

Varughese, K. I., N. H. Xuong, P. M. Kiefer, D. A. Matthews, and J. M. Whitely. 1994. "Structure and mechanistic characteristics of dihydropteridine reductase: A member of the Tyr-(Xaa)3-Lys-containing family of reductases and dehydrogenases." *Proceedings of the National Academy of Sciences of the USA* 91:5582–5586.

Vingron, M., and P. Argos. 1989. "A fast and sensitive multiple sequence alignment algorithm." *Computer Applications in the Biosciences* 5:115–122.

Wagner, R. A., and M. J. Fischer. 1974. "The string-to-string correction problem." *Journal of the Association for Computing Machinery* 21(1):168–173.

Wallace, C., and D. Dowe. 1994. "Intrinsic classification by MML—the Snob program." In *Proceedings of the Seventh Australian Joint Conference on Artificial Intelligence*, pp. 37–44 (Singapore: World Scientific).

Wang, J. T. L., G. W. Chirn, T. G. Marr, B. A. Shapiro, D. Shasha, and K. Zhang. 1994a. "Combinatorial pattern discovery for scientific data: Some preliminary results." In *Proceedings of the ACM SIGMOD International Conference on Management of Data*, pp. 115–125 (New York: ACM Press).

Wang, J. T. L., T. G. Marr, D. Shasha, B. A. Shapiro, and G. W. Chirn. 1994b. "Discovering active motifs in sets of related protein sequences and using them for classification." *Nucleic Acids Research* 22:2769–2775.

Wang, J. T. L., K. Zhang, K. Jeong, and D. Shasha. 1994c. "A system for approximate tree matching." *IEEE Transactions on Knowledge and Data Engineering* 6(4):559–571.

Wang, J. T. L., T. G. Marr, D. Shasha, B. A. Shapiro, G. W. Chirn, and T. Y. Lee. 1996. "Complementary classification approaches for protein sequences." *Protein Engineering* 9(5):381–386.

Wang, X., J. T. L. Wang, D. Shasha, B. A. Shapiro, S. Dikshitulu, I. Rigoutsos, and K. Zhang. 1997. "Automated discovery of active motifs in three dimensional molecules." In *Proceedings of the Third International Conference on Knowledge Discovery and Data Mining*, pp. 89–95 (Menlo Park, Calif.: AAAI Press).

Waterman, M. S. 1984. "General methods of sequence comparison." *Bulletin of Mathematical Biology* 46(4):473–500.

Waterman, M. S., ed. 1989. *Mathematical Methods for DNA Sequence Analysis.* Boca Raton, Fla.: CRC Press.

Waterman, M. S. 1995. *Introduction to Computational Biology: Maps, Sequences and Genomes.* Boca Raton, Fla.: Chapman & Hall.

Watson, J. D., N. H. Hopkins, J. W. Roberts, J. Argetsinger Steitz, and A. M. Weiner. 1987. *Molecular Biology of the Gene,* 4th ed. Redwood City, Calif.: Benjamin/Cummings.

Weissman, R. F. E. 1989. "In search of the scholar's workstation: Recent trends and software challenges." *Academic Computing* 3:28–31.

Wiederhold, G. 1992. "Mediators in the architecture of future information systems." *IEEE Computer* 25:38–49.

Wiener, J. L., and J. F. Naughton. 1994. "Bulk loading into an OODB: A performance study." Technical Report 1218, University of Wisconsin-Madison, available at `ftp.cs.wisc.edu/tech-reports/reports/tr1218.ps.Z`.

Wierenga, R. K., and W. G. J. Hol. 1983. "Predicted nucleotide binding properties of p21 protein and its cancer related variant." *Nature* 302:842–844.

Wierenga, R. K., M. C. De Maeyer, and W. G. J. Hol. 1985. "Interaction of pyrophosphate-moieties with α-helixes in dinucleotide binding proteins." *Biochemistry* 24:1346–1357.

Wierenga, R. K., P. Terpstra, and W. G. J. Hol. 1986. "Prediction of the occurrence of the ADP-binding $\beta\alpha\beta$-fold in proteins using an amino acid sequence fingerprint." *Journal of Molecular Biology* 187:101–107.

Willett, P. 1987a. "A review of chemical structure retrieval systems." *Journal of Chemometrics* 1:139–155.

Willett, P. 1987b. *Similarity and Clustering in Chemical Information Systems.* Letchworth, UK: Research Studies Press.

Willett, P. 1988. "Ranking and clustering of chemical structure databases." In *Physical Property Prediction in Organic Chemistry,* ed. C. Jochum, M. G. Hicks, and J. Sunkel, pp. 191–207 (Heidelberg: Springer-Verlag).

Woese, C. R., S. Winker, and R. R. Guttell. 1990. "Architecture for ribosomal RNA: Constraints on the sequence of tetra-loops." *Proceedings of the National Academy of Sciences of the USA* 87:8467–8471.

Wolfson, H. J. 1991. "Generalizing the generalized Hough transform." *Pattern Recognition Letters* 12(9):565–573.

Woodsmall, R. M., and D. A. Benson. 1993. "Information resources at the National Center for Biotechnology Information." *Bulletin of the Medical Library Association* 81:282–284.

Wootton, J. C., and S. Federhen. 1993. "Statistics of local complexity in amino acid sequences and sequence databases." *Computers in Chemistry* 17:149–163.

Wu, S., and U. Manber. 1992. "Fast text searching allowing errors." *Communications of the Association for Computing Machinery* 35(10):83–91.

Xu, Y., J. R. Einstein, R. J. Mural, M. Shah, and E. C. Uberbacher. 1994. "An improved system for exon recognition and gene modeling in human DNA sequences." In *Proceedings of the Second International Conference on Intelligent Systems for Molecular Biology,* pp. 376–384 (Menlo Park, Calif.: AAAI Press).

Zhang, K., D. Shasha, and J. T. L. Wang. 1994. "Approximate tree matching in the presence of variable length don't cares." *Journal of Algorithms* 16(1):33–66.

Zhang, X., and D. Waltz. 1993. "Developing hierarchical representations for protein structures: An incremental approach." In *Artificial Intelligence and Molecular Biology*, ed. L. Hunter, pp. 195–209 (Menlo Park, Calif.: AAAI Press).

Zhang, Z., B. Raghavachari, R. C. Hardison, and W. Miller. 1994. "Chaining multiple-alignment blocks." *Journal of Computational Biology* 1:217–226.

Zuker, M. 1989. "On finding all suboptimal foldings of an RNA molecule." *Science* 244:48–52.

Zuker, M. 1996. *Prediction of RNA Secondary Structure by Energy Minimization*. St. Louis, Mo.: Washington University Press.

Zuker, M., and P. Stiegler. 1981. "Optimal computer folding of large RNA sequences using thermodynamics and auxiliary information." *Nucleic Acids Research* 9:133–148.

Index